T0135744

Composition, Structure and Magneto-Mechanical Properties of Ni-Mn-Ga Magnetic Shape-Memory Alloys

vorgelegt von
Diplom-Ingenieur und Master of Science
Markus Chmielus

Von der Fakultät III – Prozesswissenschaften
der Technischen Universität Berlin
zur Erlangung des akademischen Grades

Doktor der Ingenieurswissenschaften
Dr.-Ing.

genehmigte Dissertation

Promotionsausschuss:

Vorsitzender: Prof. Dr. rer. nat. H. Schubert
Berichter: Prof. Dr. rer. nat. W. Reimers
Berichter: Prof. Dr. sc. techn. P. Müllner

Tag der wissenschaftlichen Aussprache: 14.06.2010

Berlin 2010
D83

Bibliografische Information der Deutschen Nationalbibliothek

Die Deutsche Nationalbibliothek verzeichnet diese Publikation in der
Deutschen Nationalbibliografie; detaillierte bibliografische Daten sind
im Internet über http://dnb.d-nb.de abrufbar.

ISBN 978-3-8325-2531-6

Logos Verlag Berlin GmbH
Comeniushof, Gubener Str. 47,
10243 Berlin
Tel.: +49 (0)30 42 85 10 90
Fax: +49 (0)30 42 85 10 92
INTERNET: http://www.logos-verlag.de

ACKNOWLEDGEMENT

I would like to thank my advisors Prof. P. Müllner, Prof. W. Reimers and Dr. R. Schneider, for their support, ideas, guidance, and organizational help during the last three years. I also want to thank Prof. Schubert for taking the chair of the dissertation defense committee. During the course of this delocalized Ph.D. program, I had the honor to work with numerous colleagues and collaborators in Berlin at the Helmholtz Centre for Materials and Energy (HZB), at Boise State University, at Northwestern University and at the Technical University of Braunschweig. I want to thank especially my colleague Katharina Rolfs for her support, sample preparation and very fruitful discussions, as well as Dr. R. Wimpory, Mirko Boin, Dr. J.-U. Hoffmann and Dr. I. Glavatskyy of the HZB and Arno Mecklenburg of MSM Krystall our collaborators at the Technical University of Braunschweig Jan Guldbakke, and Prof. A. Raatz. At Boise State University, I want to thank my colleagues Adrian Rothenbühler, Brittany Muntifering, Brittany Siewert, Cassie Witherspoon, Dave Carpenter, Dave Schenker, Doug Kellis, Kimo Wilson, Matt Reinhold, Mike Hagler, Dr. Paul Lindquist, and Nikki Kucza for their great help, ideas, and enthusiasm. The design and fabrication of upgrades and new experiments would not have been possible without the help of Phil Boysen (BSU). I want further thank Prof. D. Dunand (Northwestern University, NU), Prof. X. Zhang (Harbin Institut of Technology, China), and Peiqi Zheng (NU) for their collaboration on MSMA foams as well as Dr. S. Vogel and Dr. D. Brown of the Los Alamos Neutron Scattering Facility for their support during beamtime and data evaluation. Without samples of Prof. Kostorz (ETH Zürich) and Prof. Chernenko (Universidad del País Vasco, Bilbao) comparisons with samples of other research groups would have not been possible. I also want to express my gratitude to Prof. D. Butt (BSU), Prof. M. Frary (BSU), Prof. B. Knowlton (BSU) and their students for their help with sample preparation and characterization, as well as Dr. N. Kardilov, A. Hilger and A. Paulke for their support with x-ray tomography experiments and reconstruction work. I also want to thank Dr. K. Ullako for his interest and inspiration, and Prof. R.C. Pond, and Prof. B. Schönfeld for their discussions regarding neutron diffraction experiments and analysis. I also want to express my gratitude to the German Research Foundation priority program SPP1239 which partially funded this project, the generous support of the HZB institutes G-I1 and M-I1 and their heads Dr. K.Habicht and Prof. Dr. A. Tennant as well as funding from the U.S. National Science Foundation and the Department of Energy, Office of Basic Energy Studies.

Finally, I want to thank my friends, my family, parents and siblings and especially my wife Jennifer for their continuous support and confidence.

Boise, ID, May 2010   Markus Chmielus

ABSTRACT

Magnetic shape-memory alloys (MSMAs) are smart materials which show in single crystalline form a magnetic field induced plastic and recoverable deformation of up to 10%. Ni-Mn-Ga is the as most prominent representative. The shape change of MSMAs is based on the motion of twin boundaries driven by a magneto-stress due to an applied magnetic field. The plastic deformation takes place in the martensite phase and does not require a phase change as needed in shape-memory alloys (SMAs). The combination of high strain of SMAs and high actuation frequencies positions MSMAs as attractive smart actuator materials. A challenge of MSMAs is that magneto-mechanical properties are still very difficult to predict and to reproduce. The production of Ni-Mn-Ga single crystals is rather difficult and time consuming. Chemical segregation leads to a continuous variation of the Mn concentration in single crystals in growth direction. The composition change has a strong influence on all properties of MSMA. Furthermore, sample preparation influences magneto-mechanical properties.

This dissertation is therefore divided in three parts: first, the characterization of composition, structure, transformation temperatures, magnetic and mechanical properties as a study on position within an ingot. Second, the influence of surface polishing and surface deformation on the twinning stress. Third, the influence of training and constraints on magneto-mechanical properties.

This study demonstrates that MSMA properties depend on the position within a single crystal ingot. With increasing Mn content the martensite structure changes from 10M over 14M to nonmodulated martensites. The decrease of surface roughness leads to a decrease of twinning stress. On the other hand, polished samples have only very few twin boundaries moving rapidly which results in serrated stress-strain curves. Furthermore, surface deformations pin twin boundaries and lead to dense twin microstructures. MSMAs with twinning stresses of above 1 MPa and few twin boundaries moving through the sample only show a magnetic field-induced strain when tilting of the sample is not restricted by constraints. In soft MSMAs, tilting is not necessary since multiple twin boundaries move in different orientations. Therefore, soft samples can adapt to constraints much better than harder MSMAs and show large magnetic field-induced strain.

ZUSAMMENFASSUNG

Magnetische Formgedächtnislegierungen (MSMA) mit Ni-Mn-Ga als prominenstensten Vertreter sind faszinierende Materialien, die eine magnetfeldinduzierte plastische Verformung von bis zu 10% in einkristalliner Form aufweisen. Die Verformung der MSMA basiert dabei auf der Bewegung von Zwillingsgrenzen, die durch magnetfeldinduzierte interne Spannungen angetrieben werden. Die Verformung benötigt also nicht wie bei traditionellen Formgedächtnis-legierungen (SMA) eine Phasenumwandlung sondern findet im der Martensit Phase statt. Aus diesem Grund vereinen MSMA die hohe plastische Verformung von SMA mit den schnellen Verformungsfrequenzen und bilden somit eine attraktive Alternative zu etablierten aktiven Materialien. Die magneto-mechanischen Eigenschaften dieser Materialien sind immernoch nicht klar zu reproduzieren. Die Herstellung von Ni-Mn-Ga Einkristallen ist schwierig und zeitaufwändig. Wegen chemischer Segregation verändert sich die Zusammensetzung von Ni-Mn-Ga Einkristallen kontinuierlich in Wachstumsrichtung und mit ihr alle MSMA Eigenschaften. Ausserdem haben Probenbearbeitung und Training der Proben einen entscheidenen Einfluss auf die magneto-mechanischen Eigenschaften von MSMA.

In dieser Doktorarbeit werden aufeinander aufbauende Themen bearbeitet. Im ersten Teil werden Zusammensetzung, Struktur, Phasenumwandlungstempera-turen, magnetische und mechanische Eigenschaften von Proben aus drei Einkristallen in Bezug auf deren Position im Kristall ausgewertet. Im zweiten Teil wird der Einfluss von Oberflächenrauhigkeiten und –verformungen auf die Zwillingsspannung von MSMA untersucht. Im dritten Teil wird der Einfluss von Training und Einspannungen auf das magneto-mechanische Verhalten analysiert werden. Die Untersuchungen dieser Dissertation haben den Zusammenhang von MSMA Eigenschaften mit Bezug auf die Position und damit auf diese chemische Zusammensetzung herausgestellt. Als besonders wichtiges Resultat ist hier die Veränderung der Martensitstruktur von 10M über 14M zu nichtmoduliertem Martensite mit ansteigendem Mn Gehalt genannt. Die Verringerung der Oberflächenrauhigkeit verringert auch die Zwillingsspannungen in MSMA Kristallen. Die Anzahl von Zwillinggrenzen ist in polierten Proben geringer und ausserdem bewegen sich die Zwillinggrenzen über größere Strecken als in unpolierten Proben. Im letzten Teil der Arbeit wurde festgestellt, dass MSMA mit Zwillingsspannungen von mehr als 1 MPa nur dann magnetfeldinduzierte Dehnung aufweisen, wenn die Probe während der Verformung eine Kippbewegung ausführen kann. Falls Einspannungen dies verhindern, ist keine magnetfeldinduzierte Dehnung messbar. In MSMA mit Zwillinggrenzspannungen unter 1 MPa spielen diese Einspannungbedingungen keine Rolle, da Zwillinge in verschiedenen Bereichen gebildet werden und sich den Einspannung besser anpassen als in harten MSMA.

TABLE OF CONTENT

1. INTRODUCTION AND MOTIVATION

Between the report of Friedrich Heusler in 1901 on alloys with an $L2_1$ structure today known as Heusler structure and parent structure of Ni-Mn-Ga magnetic-shape-memory alloys (MSMAs), and the first report of magnetic field-induced strain (MFIS) in 1996, nearly one century passed. The properties of MSMAs, especially their MFIS of up to 10% has since attracted numerous scientists. Since 2006, the German research foundation funded 28 research groups over a six year period in the SPP 1239 priority program in all areas of MSMA research from theory over thin film to bulk materials.

Over a decade of research, the study of magneto-mechanical properties especially high cycle properties were limited to case studies. The properties were not predictable and especially not designable. Therefore, in 2007 Dr. Rainer Schneider (Helmholtz Centre Berlin for Materials and Energy) with new single crystal growth capabilities and Prof. Peter Müllner (Boise State University) with unique magneto-mechanical testing capabilities, created a collaborative effort with a delocalized Ph.D. position in conjunction with the Technical University Berlin and Prof. Walter Reimers as "Doktorvater". The combination of expertise was the base for a systematic study on MSMA properties especially magneto-mechanical properties. In this study, the influencing parameters should not be restricted to a few samples and in a qualitative manner, but on many samples, systematically and quantitatively. Therefore, all important MSMA properties, i.e. thermal, structural, magnetic, and mechanical properties, should be analyzed over the entire height of the supplied single crystal ingots. Furthermore, the larger amount of samples, should have been sufficient to deconvolute influencing parameters, e.g. surface preparation and different types of training, and thus distinctively identify the impact of different factors on sample preparation and training.

Finally, the results of this study should outline new strategies to move MSMAs closer towards technological applications including actuators, self-powering sensors, and micro energy harvesting devices.

2. CONCEPTUAL FORMULATION

The work in this dissertation is directed towards three consecutive areas of research. First, all important MSMA properties, i.e. chemical composition, martensitic phase transformation temperature, Curie temperature, lattice parameters, martensite structure, saturation magnetization, magnetic anisotropy, and twinning stress, are to be characterized as a function the position within Ni-Mn-Ga single crystal ingots. Other experiments have to be upgraded or new additions to already existing experiments have to be designed and built, e.g. a clamping mechanism for magnetic anisotropy measurements in the VSM.

In a second part, the influence of surface treatments and surface deformations on twinning stresses of single crystalline Ni-Mn-Ga samples is to be analyzed. For this series of experiments, unpolished samples are polished in several subsequent stages and the surfaces of polished samples are mechanically deformed with different methods. After every subsequent electropolishing or surface deformation step, the twinning stress is are measured. In parallel, the twinning stress of unpolished reference samples is tested in the same way to deconvolute the influence of surface treatment from the influence of repeated twinning stress tests (i.e. mechanical training).

Finally, the influence of different training methods on the high cycle magneto-mechanical properties of Ni-Mn-Ga single crystals is to be investigated. For this part of the dissertation, optical upgrades for the high cycle test as well as an optical-magneto-mechanical device to observe deformation behavior with an applied magnetic field and under different constraints are designed and built.

This dissertation presents a systematic study of the influence of composition and martensite structure on twinning stress and magnetic field-induced strain of Ni-Mn-Ga single crystalline samples.

3. BACKGROUND

3.1. History

In 1901, the German mining engineer and chemist, Friedrich Heusler (1866-1947), submitted a note to the Deutsche Physikalische Gesellschaft e.V. (German Physics Society) about alloys, which are ferromagnetic even though their components are not [1]. This note was published 1903 together with two additional reports about ferromagnetic Mn alloys by F. Heusler and two of his colleagues W. Starck and E. Haupt [2,3]. With the discovery of these so called "Heusler alloys" (see section 3.4), the history of magnetic shape-memory alloys, of which many are Heusler alloys, began. Later, in 1968 [4], one of the most characteristic behaviors of MSMA namely permanent plastic deformation induced by a magnetic field, was reported by Rhyne. The plastic deformation was discovered by him during experiments with very large magnetic fields (up to 10 T) and at cryogenic temperatures (4K). This "magneto-plastic" behavior of Dy was examined in more detail in 1976 [5].

The first report on magnetic-field-induced strain in MSMA was given in 1996 by Ullakko [6]. The strain was 0.19% and was measured on a Ni_2MnGa sample at 265 K in a magnetic field of 0.43 T. From then on, the interest in magnetic shape memory alloys grew rapidly all over the world.

Researchers from Spain [7], Ukraine [8], Canada [9], and China [10] examined the crystal structure of nonstoichiometric Ni_2MnGa. Murray [11] achieved a 6% magnetic-field-induced strain in nonstoichiometric Ni_2MnGa at room temperature in a magnetic field of 0.62 T while different stresses were applied to restore the magnetic-field-induced elongation of the sample and to measure the magnetic-field-induced stress. In 2002, Sozinov [12] published results of a magnetic-field-induced strain of 9.5% of nonstoichiometric Ni_2MnGa in a 1.05 T magnetic field at room temperature. This strain was measured in a unidirectional magnetic field.

Starting in 2002, Müllner published results of experiments with nonstoichiometric Ni_2MnGa that received different heat treatments and were tested over several cycles in a rotating magnetic field [13-15]. Depending on the details of the heat treatment, some tested specimens showed high strains (9.6%) over a short amount of cycles, other samples exhibited strains of less than 1% over 27 million magneto-mechanical cycles.

In 2006, the German Research Foundation (DFG) started the SPP1239 priority program. This priority program concerts and funds the research on MSMA of 24 to 28 (depending on the funding period) research groups. Divided in three areas, fundamentals of the materials (area A), production of bulk material and development of applications of bulk material (area B), and thin film and their applications (area C) are examined in numerous groups within the German research landscape [16]. The present dissertation was partially funded within the SPP1239 area B.

Due to the high cost of single crystal Ni-Mn-Ga bulk production, research was also directed towards polycrystalline bulk materials and thin films, as seen in the SPP1239 program [16] and in numerous publications, i.e. [17,18]. Even though fine-grained polycrystalline MSMA are easier to produce than single crystalline bulk MSMA, their net magnetic-field-induced strain is nearly zero due to grain boundaries hindering twin boundary motion [19-21]. By producing a porous material, i.e. foam, grain boundaries can be removed and if grains are spanning between these pores, twin boundaries can move as freely as in single crystalline bulk material within the grains. Combining the properties of single crystalline high magnetic-field-induced strain and lower manufacturing costs of polycrystalline Ni-Mn-Ga, the first work on magnetic shape memory alloy foam has been conducted and magnetic-field-induced strain of 0.12% was published [22,23] by the research groups of Müllner and Dunand. In 2009 thermo-magneto-mechanical cycling was used to further train the magnetic shape-memory alloy foams to reach magnetic-field-induced strains of 8.9% [24,25].

Since 1996, when the MSMA effect was discovered on off-stoichiometric Ni_2MnGa, many different compositions have been investigated, i.e. Ni-Fe-Ga [26], Ni-Mn-In [27], Co-Ni-Al [28], Co-Ni-Ga [29]. Furthermore, Ni-Mn-Ga was alloyed with a forth element i.e. Co [30], Fe [31-33], Al, and rare earth metals [34], to improve certain characteristics of standard Ni-Mn-Ga MSMA. In this dissertation, the most common and also commercially available MSMA namely Ni-Mn-Ga in single crystalline form was investigated.

3.2. PHENOMENOLOGY

The most distinguishing property of MSMA is the macroscopic change of shape when exposed to a magnetic field. The shape change though is to be divided into magnetoelastic and magnetoplastic deformation [35]. During magnetoelastic deformation the initial shape is restored when the applied magnetic field is removed. During magnetoplastic deformation the shape change is reversible when turning the magnetic field direction by 90° and, alternatively, when applying a stress perpendicular to the direction of the magnetic field, which caused the shape change. In this work, only the magnetoplastic deformation of MSMA is examined. A macroscopic plastic deformation of MSMA can also be achieved by applying mechanical stress. If the stress is applied perpendicularly to a magnetic field, the deformation is recovered (see Figure 3.1, curves labeled "2 T"). If the stress is applied on a sample without an applied magnetic field, the deformation is not recovered (see Figure 3.1, curves labeled "0 T"). In Figure 3.1 it can also be seen that an applied magnetic field increases the stress needed to deform sample. The stress difference at similar strains is called magneto-stress $\Delta\sigma_M$. The strain resulted from the macroscopic shape change by the magnetic field is called magnetic-field-induced strain (MFIS). Independent of the stress needed to initiate twin boundary motion (twinning stress), there is normally a threshold stress level below which twins are not moving followed by a twinning stress plateau. This threshold stress originates from the attractive forces between disconnections, which has to be overcome by an applied stress. This threshold stress and stress plateau was also observed by numerical simulations of twin-twin interaction [36,37].

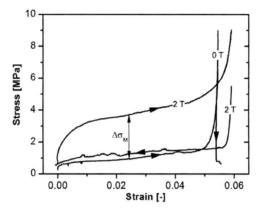

Figure 3.1: Macroscopic shape change of MSMAs in an orthogonal magnetic field. Curves labeled "2 T" show the stress-strain curve of a MSMA in a magnetic field of 2 T and curves labeled "0 T" without magnetic field. If the MSMA is plastically deformed without applied magnetic field, the deformation is not recovered.

Figure 3.2 shows the MFIS as a function of variable magnetic field strength and is based on results by Müllner et al. [35]. When the magnetic field is turned on for the first time, the compressive strain increased to a maximum of -1.67%[1] at a magnetic field of around 0.8 T. By further increasing the magnetic field, the strain does not increase further. When the magnetic field is turned off and the direction is reversed, the strain remains nearly constant (Figure 3.2). The MFIS is actually reduced slightly by 0.02% to -1.65%, which is caused by elastic relaxation (see inset [35]). With increasing magnetic field, the MFIS is increasing to -1.67% again. The maximum MFIS is reached at a field at approximately 1.2 T. With decreasing magnetic field, the maximum magnetic-field-induced strain stays constant to about 0.5 T and reduces by 0.02% below 0.5 T.

---

[1] The negative sign indicates that the strain is compressive. Mathematical convention implies that at -1.62% there is a minimum. Following the convention used in the MSMA literature, we refer to a "maximum" since this strain value represents the largest length change.

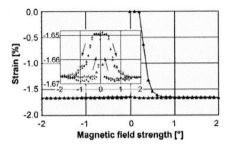

Magnetic field strength [°]

Figure 3.2: Results of magneto-mechanical experiments of a magnetoplastic MSMA with constant magnetic field direction and changing magnetic field strength. After the first increase of the magnetic field the strain reaches -1.67% and is with decreasing and reversing magnetic field nearly constant. An elastic relaxation of 0.02% is present at low magnetic fields seen in the inset. The arrows indicate the loading and unloading direction. Reprinted with permission of TMS and P. Müllner, [35], © The Minerals, Metals & Materials Society (TMS), 2005.

When MSMA are exposed to a rotating magnetic field, their macroscopic shape is changing periodically with the magnetic field direction (see [13-15] as shown in Figure 3.3. During each full rotation of the magnetic field, the MSMA is expanding and shrinking in one direction twice. Each set of expansion and shrinking is called magneto-mechanical cycle and is exemplarily shown in Figure 3.3a. Thus, for each rotation of the magnetic field, a MSMA is exhibiting two magneto-mechanical cycles. In Figure 3.3, it can also be seen how training is affecting the strain-magnetic-field-direction curves. Effectively trained MSMA tend to have large MFIS and a curve with squared form, while ineffectively trained MSMA tend to have lower strain and round curves [35].

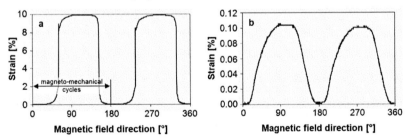

Figure 3.3: Results of rotating field experiments. The periodic MFIS of an effectively trained MSMA (a) is large and has a step-like change, while the MFIS of an ineffectively trained MSMA is two orders of magnitude smaller (b). Reprinted with permission of TMS and P. Müllner, [35], © The Minerals, Metals & Materials Society (TMS), 2005.

Details of MFIS behavior strongly depend on the composition of the MSMA. The most commonly used MSMA are nonstoichiometric Ni_2MnGa alloys. Due to the sensitivity of the martensite phase to the temperature of the sample, the operating temperature at which experiments are performed is also critical. This is the case because MSMA show the magnetic-field-induced strain only in the martensitic phase and not in the austenitic parent phase. The temperature range, in which the martensitic transformation takes place, strongly depends on the composition of the alloy. For stoichiometric Ni_2MnGa, the martensitic phase transformation is below room temperature and therefore the martensitic phase not present at room temperature. By varying the composition, the martensitic phase transformation temperature can be raised above room temperature. Depending on composition and temperature, Ni-Mn-Ga MSMA can be in different phases. Most importantly for applications and most commonly investigated are the 10M and 14M (M stands for "modulated") martensites, which exhibit theoretically up to 6% and 10% MFIS, respectively. The MFIS also depends strongly on the training (see section 4.5). In well trained and nearly single variant MSMA, MFIS as large as 9.6% have been reported [14]. Samples with large initial strain and only one twin variant fail after a few thousand cycles. Samples with small initial strain and several twin variants present last longer than 100 million magneto-mechanical cycles [15,38].

3.3. MAGNETISM

3.3.1. INTRODUCTION AND UNITS

As for all areas of science and engineering, it is most important to use units internationally accepted and agreed on. While this common rule seems to be obeyed for most areas of research at least when it comes to publishing results in peer reviewed journals, this is not the case for magnetism. In the field of magnetism, even today, Gaussian units are in use next to SI units. This leads to unnecessary confusion and errors especially since most manufacturers for magnetic test devices seem to use Gaussian units as data output. For this reason, a short table of most common units and conversion between SI and non SI units is given here [39].

Table 3.1: SI and Gaussian units and conversion between them in magnetism [39].

Quantity	Symbol	Gaussian System	Conversion Factor	SI System
Magnetic flux density	$\boldsymbol{B} = \mu_0\boldsymbol{H}$	G	10^{-4}	T
Magnetic field strength	H	Oe	$1000/4\pi$	A/m
Volume magnetization	M_V	emu/cm^3	1000	A/m
Mass magnetization	M_m	emu/g	1	A m^2/kg, J/T kg
Magnetic moment	m_{mag}	emu	10^{-3}	A m^2, J/T
Volume susceptibility	χ	dimensionless	4π	dimensionless
Permeability, $\mu = \mu_0\mu_r$	μ	$\mu^{\cdot} = \mu_r$	$4\pi \times 10^{-7}$	H/m, N/A^2
Relative permeability	μ_r	dimensionless	1	Dimensionless
Magnetic constant	μ_0	not used, $\mu^{\cdot} = \mu_r$	-	H/m, N/A^2
Demagnetization factor	N_D	dimensionless	$1/4\pi$	dimensionless

An important constant in magnetism is the magnetic constant or permeability of free space, μ_0, which is $4\pi \times 10^{-7}$ N/A^2. The conversion between measured values usually in Gaussian units and the reported results need to be performed very carefully.

Ferromagnetism is one of the important properties of MSMA. All materials can be divided into five groups according to their magnetic properties: diamagnetic, paramagnetic, ferromagnetic, antiferromagnetic, and ferrimagnetic. The volume magnetization, M_V, is the volume density of magnetic moments, m_{mag}, while the mass magnetization, M_m, is the mass density of the magnetic moments with:

$$M_V = \frac{1}{V}\sum_{i=1}^{N} m_{mag} \text{ and } M_m = \frac{1}{V}\sum_{i=1}^{N} m_{mag} \tag{3.1}$$

The magnetization is related to the applied magnetic field, H, through the magnetic susceptibility χ. The magnetic susceptibility is therefore a measure of the magnetization of a material in response to an applied magnetic field:

$$M_V = \chi H \tag{3.2}$$

The magnetic flux density, B, in free space is the magnetic field strength, H, times the permeability of free space, μ_0:

$$B = \mu_0 H \tag{3.3}$$

The demagnetization factor, N_D, is a factor based on the geometry of a sample and takes into account that there is a demagnetizing field within a magnet that opposes the external magnetizing field. Depending on form of the sample, the magnetization within the sample depends on its orientation towards the magnetic field. Therefore, the applied magnetic field, H_{appl}, has to be converted to the effective magnetic field, H_{eff}, using the shape of the sample as shown in [40]. The effective magnetic field is related to the applied magnetic field as followed:

$$H_{eff} = H_{appl} - N_D M \tag{3.4}$$

with N_D as the demagnetization factor and M the magnetization of the sample. N_D is calculated as followed:

$$N_D = 2(3q - 1) / (4\pi q^2) \tag{3.5}$$

with q as a geometry factor equals the length of the sample parallel to the magnetic field direction, m_{length}, over the square root of the cross section normal to the field direction, s_{mag}:

$$q = m_{length} / s_{mag} \tag{3.6}$$

The demagnetization is explained in further detail in section 4.2.3. More elaborate treatments of magnetism can be found in [41-44].

3.3.2. DIAMAGNETISM

In diamagnetic materials there are no unpaired electrons, which would contribute to a magnetic moment, are not present. Therefore, atoms of diamagnetic materials have no magnetic moment (Figure 3.4a). Their magnetic susceptibility is small and negative (Figure 3.4d) and is independent of temperature. The slightly negative susceptibility is caused by an induction phenomenon. According to Lenz's law the induction reduces the applied magnetic field in the material (Figure 3.4c). Examples for diamagnetic materials include Ga, Si, Au, and Ag.

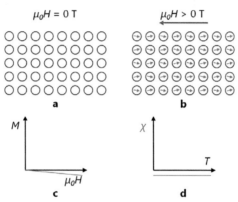

Figure 3.4: Schematic properties of diamagnetic materials. Without a magnetic field there are no magnetic moments (a). With applied magnetic field the induced magnetic moments are in opposite direction of H (b). A negative susceptibility can be observed (c). The magnetic susceptibility remains constant over temperature (d). See [39,43].

3.3.3. PARAMAGNETISM

In paramagnetic materials, atoms have a magnetic moment (Figure 3.5a). Due to thermal energy, the magnetic moments are randomly orientated. With increasing external magnetic field the randomly orientated magnetic moments become more and more aligned parallel to the magnetic field direction (Figure 3.5b). The magnetic susceptibility is small and positive (Figure 3.5c) and is temperature dependent (Curie law). At very high magnetic fields, the magnetization saturates. With increasing temperature, it becomes more difficult to align the magnetic moments because of higher thermal energy. An example for a paramagnetic material is Pt.

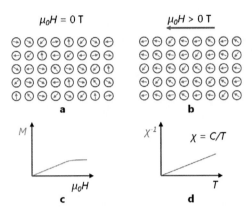

Figure 3.5: Schematic properties of paramagnetic materials. The arrangement of magnetic moment without external magnetic field is shown in (a) and with magnetic field (b), in which the magnetic moments becoming aligned to the magnetic field. The magnetic moment depending on a magnetic field shows a positive susceptibility and a saturation behavior at high magnetic fields (c). The magnetic susceptibility decreases with increasing temperature. C is the Curie constant. See [39,43].

3.3.4. FERROMAGNETISM

Ferromagnetic materials demonstrate a very distinct magnetic behavior. When ferromagnetic materials are consecutively magnetized in opposing directions, the magnetization of the material shows a hysteresis. A schematic of this hysteresis is shown in Figure 3.7. When the ferromagnetic sample is first magnetized (virgin curve) in one direction (a), the magnetization increases until a saturation magnetization, M_{sat}, is reached at the saturation field, H_{sat}, at point (b). Upon decreasing the magnetic field to zero, the magnetization decreases as well, but not to zero. A residual magnetization remains at zero magnetic field, which is called (positive) remanence, $+M_r$. When the direction of the magnetic field is changed and increased, the magnetization of the sample decreases to zero at the so called (negative) coercive field strength, $-H_c$ (d). Upon further increase of the magnetic field in negative direction the (negative) saturation magnetization, $-M_{sat}$, is reached (e). Decreasing and reversing the magnetic field again, first decreases the magnetization of the sample until it reaches at a field strength of zero the (negative) remanence, $-M_r$ (f). At a positive magnetic field or (positive) coercive field strength, $+H_c$ (g), a neutral magnetization is reached and with further increasing magnetic field strength the (positive) saturation magnetization is reached again at (b).

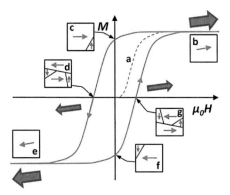

Figure 3.6: Magnetization hysteresis of ferromagnetic materials. (a) is the virgin curve. When the magnetization flattens out, the saturation magnetization, M_{sat}, is reached at (b). Point (b) shows the positive remanence, $+M_r$, and point (f) $-M_r$. The points (d) and (g) show the negative coercive field strength, $-H_c$, respectively $+H_c$ at which the magnetization is zero. At point (c) the negative saturation, $-M_{sat}$, is reached. The small square next to the indicated hysteresis points show schematically the alignment of magnetic domains due to the magnetic field. The direction and strength of the magnetic field is implied by the blue arrows. Based on [44].

Two mechanisms are causing the specific magnetic behavior of the magnetization of ferromagnetic materials: the motion of magnetic domain walls and the rotation of magnetic moments. Magnetic domains are small volume regions with a mutual alignment of their magnetic moments. They are separated from each other by magnetic domain walls, in which the magnetic dipoles are gradually changing their direction from one to the other domain moment direction. The magnetic domains and magnetic domain wall motion is the cause for the hysteresis of the magnetization curve. When magnetized in a certain direction, domains with a magnetic moment more or less parallel to the applied magnetic field grow by domain wall motion on the expense of domains with less parallel magnetic moments. Upon further increase of the external magnetic field, the magnetic moments of the favored magnetic domain if slightly misaligned will rotate until completely parallel to the magnetic field (Figure 3.7b). A decrease of the magnetic field causes a realignment of the magnetic moments towards the magnetization direction of the magnetic domain. When further decreasing the magnetic field, some magnetic domains eventually grow that are not aligned with the magnetic field (d). When reversing the magnetic field direction, domains with magnetic moments parallel to the magnetic field are growing until the net magnetization of the sample is zero. Further magnetic field increase results again in growth of preferred magnetic domains and finally rotation of magnetic moments as described above.

The magnetic moments of a ferromagnetic domain are aligned as shown in Figure 3.7a. In Figure 3.7b it is shown, that the saturation magnetization, M_{sat}, of ferromagnetic materials is decreasing with increasing temperature. Due to thermal energy, the alignment of magnetic moments decreases and gets lost above the Curie temperature of the material. The inverse magnetic susceptibility is increasing above the Curie temperature, which is called Curie-Weiss law. Examples of ferromagnetic elements are Fe, Co, Ni, and Dy.

The ferromagnetic phase transformation that is taking place at the Curie temperature is a second order phase transformation. The characteristic of a second-order phase transformation following the Ehrenfest classification is the continuity of the first derivation of the Gibbs free energy but discontinuity of the second-order derivation. Using modern classifications, the ferromagnetic phase transformation is also a second-order transformation since there is no latent heat associated with this phase transformation.

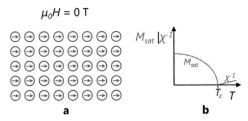

Figure 3.7: Schematic properties of ferromagnetic materials. Without magnetic field the magnetic moments are aligned (a). The temperature dependence of the magnetization of ferromagnetic materials is given in (b). In a saturating magnetic field, the magnetization is decreasing from e.g. the saturation magnetization, M_{sat}, to zero at the Curie temperature, T_c, of the material. Based on [39,43].

3.3.5. OTHER TYPES OF MAGNETISM

Antiferromagnetism and ferrimagnetism are other forms of magnetism with strongly coupled magnetic moments. In antiferromagnetic materials, the direction of magnetic moments alternates resulting in zero net magnetization. The magnetic susceptibility of antiferromagnetic materials is small and positive. Mn is an antiferromagnetic element. Atoms of ferrimagnetic materials have magnetic moments that have alternating directions but do not cancel each other out. This leads to a net magnetic moment. Their magnetic susceptibility is large and positive.

3.3.6. MAGNETIC ANISOTROPY

Figure 3.6 shows the magnetization hysteresis of a ferromagnetic material. The form of the hysteresis curve and the saturation magnetization depends on several factors, including temperature, crystal structure, magnetization direction, and constituents. In single crystals the slope of the M-H curve to magnetic saturation depends on the crystallographic direction. The difference of the M-H curves in single crystals is called magnetic anisotropy. Figure 3.8 shows the magnetization curves for Ni and iron magnetized along the crystallographic directions [100], [110], and [111]. For iron and Ni, there is one direction where the saturation magnetization is reached with a lower magnetic field than in the other directions. This direction is called the direction of easy magnetization. This direction is [100] for iron, [111] for Ni, and [001] for Ni_2MnGa (see section 3.4.3, using the pseudo-orthorhombic and pseudo-tetragonal axes system). The area between the magnetization curves and the magnetization axis indicates the energy needed to reach the saturation magnetization. This is exemplarily shown for the [100] direction of iron in Figure 3.8 by the hatched area. The difference between energy needed to magnetize the sample completely in the easy and hard magnetization direction is called magnetic anisotropy energy. The magnetization energy for all crystal directions can be illustrated as a three dimensional surface with the lowest magnetization energy in the easy and the highest in the hard magnetization direction.

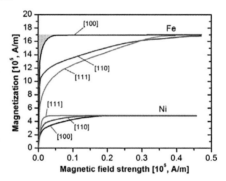

Figure 3.8: Magnetization curves for single crystals of iron (bcc) and Ni (fcc). For both materials, the form of the M-H curve and the magnetic field need to fully saturate the sample depends on the crystallographic direction. The directions of easy magnetization are [100] for iron and [111] for Ni. It is more difficult to reach the saturation magnetization for other directions. The hatched area between the magnetization curves and the magnetization axis indicate the energy needed to reach the saturation magnetization in the [100] direction for iron. Based on graphs of [45,46].

3.4. HEUSLER ALLOYS

Materials - later called Heusler Alloys - were first described in 1903 [1] by Friedrich Heusler (1866-1947), a German mining engineer and chemist. He examined very remarkable properties of Cu_2MnAl. While none of its compounds shows ferromagnetic properties, Cu_2MnAl is ferromagnetic. This early definition of Heusler alloys has changed. Today, Heusler alloys are defined to be ferromagnetic materials and exhibit the Heusler phase (fcc $L2_1$-ordered cubic phase). Ni_2MnGa, the most investigated MSMA, is ferromagnetic in the $L2_1$ phase below the Curie temperature. Despite the early definition of Heusler alloys, the compound Ni is ferromagnetic. This section describes the special properties of Heusler alloys in particular of Ni_2MnGa.

3.4.1. STRUCTURE OF HEUSLER ALLOYS

Heusler alloys have a $L2_1$-ordered fcc crystal structure (space group $Fm\bar{3}m$ (225), see [47,48]) with an X_2YZ composition. X and Y are mostly transition metals. Y can also be a rare earth metal [49]. Z is a non-metal or a non-magnetic metal. From [48] the following atomic positions are given in Table 3.2:

Table 3.2: Atomic positions for X, Y, and Z. X occupies eight positions, Y and Z each four positions.

X	Y	Z
[¼, ¼, ¼], [¾, ¼, ¼], [¼, ¾, ¼],	[½, 0, 0], [0, ½, 0],	[0, 0, 0], [½, ½, 0],
[¼, ¼, ¾], [¾, ¾, ¼], [¾, ¼, ¾],	[0, 0, ½], [½,½,½]	[½, 0, ½], [0, ½, ½]
[¼, ¾, ¾], [¾, ¾, ¾]		

Figure 3.9 illustrates the X_2YZ crystal structure of the Heusler phase. X occupies the orange positions, Y the blue positions, and Z the grey positions. For Ni_2MnGa which is shown in Figure 3.9 Ni is X (orange), Mn is Y (blue), and Ga Z (grey).

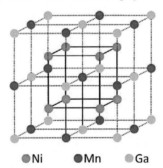

● Ni ● Mn ● Ga

Figure 3.9: Crystal structure of the Heusler phase, here for Ni_2MnGa, modified from [48]. Ni is occupying the orange sites, Mn the blue sites, and Ga the grey sites of the unit cell. Figure is based on [50].

3.4.2. PHASE TRANSFORMATIONS OF HEUSLER ALLOYS

3.4.2.1. MARTENSITIC PHASE TRANSFORMATION

The martensitic phase transformation is a diffusionless phase transformation in metal, e.g. steel, shape memory alloys, between the austenite phase and the metastable martensite. While no long-range diffusion of atoms occurs during the martensitic phase transformation, a rather subtle change of the position of atoms in the crystal lattice results in a tetragonal distortion of the cubic austenite unit cell. This small distortion is one of the basic properties of the materials class of shape memory alloys [51].

MSMA are a special case of shape memory alloys and therefore exhibit the martensitic phase transformation. Heusler Alloys have two important phases: the $L2_1$-ordered cubic or austenite phase and the martensite phase. When starting in a martensite phase, austenite starts to form above the austenite start temperature A_s and is completely transformed to austenite above the austenite finish, A_f, temperature. The austenite in Heusler alloys is the also the defining Heusler phase and has a $L2_1$-ordered face-centered cubic, fcc, crystal lattice. When decreasing the temperature of a sample fully in the austenite phase, martensite forms below the martensite start temperature, M_s. The sample is fully transformed to martensite below the martensite transformation temperature, M_f.

The martensitic transformation displays a temperature hysteresis upon heating and cooling. This phase transition a first order phase transformation, due to the discontinuity of first derivation of the Gibbs free energy (Ehrenfels definition) and association of latent heat to the transformation (modern definition). Figure 3.10a shows exemplarily a differential scanning calometry (DSC) curve with a martensitic phase transformation of a Ni-Mn-Ga sample. Upon cooling (blue curve), the phase transformation starts at the martensitic start temperature M_s and ends at the martensitic finish temperature M_f. Upon heating (red curve of Figure 3.10), the reverse transformation begins at the austenitic start temperature A_s and ends at the austenitic finish temperature A_f. The phase change from the martensite to austenite is an endothermic reaction, which means that heat is consumed for this process and the enthalpy change of the system is positive, while the change from the austenite to martensite is an exothermic reaction, energy is set free and the enthalpy of the system is decreasing. Additionally to the definition of A_s, A_f, M_s, and M_f, during DSC measurements (see Figure 3.10a) the temperatures at the maximum endothermic and exothermic heat flow are defined as A_p and M_p, respectively. For DSC measurments, an average martensitic transformation temperature, T_m, is defined as $(A_P+M_P)/2=T_m$. For magnetization curves versus temperature curves, the martensitic transformation temperature is defined as $(A_s+A_f+M_s+M_f)/4=T_m$. This average martensitic transformation temperature, T_m, is very close to the equilibrium temperature T_{eq} of the two phases and is used throughout this thesis.

27

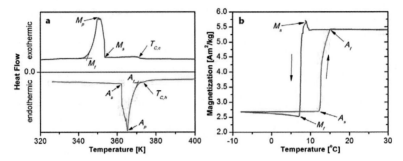

Figure 3.10: Exemplarily Differential Scanning Calometry (DSC) curve (a) and magnetization versus temperature curve (b). Heating is indicated by the red, cooling by the blue curve. Upon heating the martensitic phase starts to change to the austenitic phase at the austenitic start temperature A_s and is complete at the austenitic finish temperature A_f. Upon cooling the phase change between austenite and martensite is starting at the martensitic start temperature M_s and finished at the martensitic finish temperature M_f. In (a) A_p and M_p indicate the temperature of the peak heat flow upon heating and cooling. $T_{C,h}$ and $T_{C,c}$ mark the Curie temperatures upon heating and cooling in (a).

The hysteresis in the magnetization versus temperature curve (Figure 3.10b) can be explained by the Gibbs free energy versus temperature diagram shown in Figure 3.11. The blue arrows show the path of the Gibbs free energy upon cooling from the austenite to the martensite phase, the red arrows upon heating from the martensite to austenite phase. To form austenite when in the martensite phase, or martensite when in the austenite phase, the sample has to be overheated or undercooled. The temperature difference, ΔT_s, between the equilibrium temperature and martensite and austenite start temperature, M_s and A_s, produces a difference in Gibbs Free Energy, ΔG_{m-a} (upon heating) or ΔG_{a-m} (upon cooling), between the Gibbs free energy of austenite and martensite. Is the energy difference as large as the sum of the chemical energy needed to perform the structure change, ΔG_c, the surface energy between the austenite and martensite phase, ΔG_s, and an elastic energy term, ΔG_e, the phase change is starting (and M_s or A_s are reached). The elastic energy and surface energy are combined in most martensitic phase transformations as large as the structure change energy [51].

The austenite start and finish temperatures and the martensite start and finish temperatures depend very strongly on the composition of the alloy. For Ni_2MnGa alloys the composition dependence of the martensitic phase change was examined by [52,53]. These observations show a huge variation of M_s from 154 K to 458 K by changing the atomic percentage of the elements by 5 at.-%.

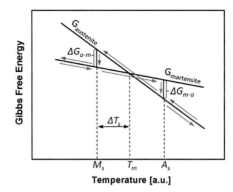

Figure 3.11: Gibbs Free Energy versus temperature during a martensitic phase transformation. When in the martensite phase, the martensite is stable until the sample receives enough energy (by overheating), so that the phase transformation strain energy, $\Delta G_{m\text{-}a}$ (sum of chemical, surface and elastic phase change energy), has been reached. At this point, austenite is starting to form. In the austenite phase upon cooling, $\Delta G_{a\text{-}m}$ has to be reached (by undercooling), so that martensite is starting to form. This is the origin of the hysteresis of the martensitic phase transformation [51].

3.4.2.2. PREMARTENSITIC TRANSFORMATIONS

Premartensitic phase transformations (PMT) have been observed in Ni-Mn-Ga MSMA with Curie temperatures well above the martensitic phase transformation temperature. Upon cooling, these samples exhibit a drastic increase of the elastic modulus above the martensitic phase transformation temperature (e.g. from 10 to 19 GPa [54]). Additionally to the increase of the elastic modulus, the internal friction is increasing from zero to a maximum in the premartensite phase before dropping again in the martensite phase [55]. This sudden soft mode phonon freezing is characteristic of the premartensite phase and can only be observed in the premartensite phase. In [54], only samples with an electron concentration of e/a < 7.6 showed the elastic modulus increase. Since soft mode phonon freezing is taking place above the martensitic phase transformation temperature and still in the fcc phase, this phase transformation is called premartensitic phase transformation.

3.4.2.3. INTERMARTENSITIC TRANSFORMATIONS

Intermartensitic phase transformations (IMT) are phase transformation from one martensite to another. IMTs and the martensite structure itself depend strongly on composition and external forces. Therefore, several different IMT have been observed, e.g. upon cooling from austenite parent phase, to a modulated martensite and further to a nonmodulated martensite phase. Upon heating, only a phase transformation from the nonmodulated martensite to austenite phase was observed [56]. Additionally, transformations upon cooling from an austenite parent to a 10M martensite and further to a 7M martensite have been observed. Depending on the martensite phase the sample was in when the cooling was stopped, 7M to austenite and 10M to austenite phase transformation have been observed, but no IMT upon heating [57]. An overview of the influence of stress and temperature on the martensite phase is given by Chernenko et al. [58]. It shows that the nonmodulated martensite is the ground state of martensites, i.e. upon cooling and with increasing stress the austenite phase changes to a 10M then 14M and then nonmodulated phase or to 14M and then nomodulate or directly from the austenite to the nomodulated phase [58]. All this, as mentioned above, strongly depends on the composition of the sample.

3.4.2.4. SELFACCOMMODATED MARTENSITE

Upon cooling from the austenite to the martensite phase, martensite variants are starting to grow at different positions of the sample. If no force or magnetic field is applied during the phase transformation, the unit cells and the twins they are building in the separate martensite variants have different orientations to accommodate the invariant plane strain between the forming martensite and the still existing austenite. Due to the invariant plane strain, slip or twinning has to occur in the martensite phase to accommodate the strain. The formation of an invariant plane by slip and twinning is schematically shown in Figure 3.12. The boundaries between equal orientated twins are called twin boundaries, which are optically visible on selfaccommodated martensites.

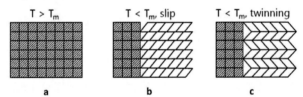

$T > T_m$	$T < T_m$, slip	$T < T_m$, twinning
a	b	c

Figure 3.12: Schematic of the interface between austenitic (striped) and martensitic phase during the martensitic phase transformation. (a) shows the crystal structure at a temperature above the martensitic phase transformation, (b) and (c) during phase transformation. Due to the constraint of the austenitic phase, slip (b) or twinning (c) is occurring in the martensitic phase to establish an invariant plane. Based on [59].

During the martensitic phase transformation, martensite grains can grow in all direction while consuming the austenite phase until they encounter another martensite grown from a different point in the sample. These neighboring martensites can now again interact with each other. All these leads to self-accommodation in the martensite phase [51], which means that strains due to the invariant plane between austenite and martensite during the phase transformation are accommodated for. Therefore, fine twins are appearing in the martensite phase if the sample was not influenced by stress or a magnetic field during the phase transformation, which happens during thermo-mechanical and thermo-magnetic training see section 4.5).

Figure 3.13 shows two different selfaccomodated twinning structures in two regions of a $(Ni_{51}Mn_{28}Ga_{21})_{99.5}Dy_{0.5}$ sample [36,37]. The topography of both regions is shown in (a) and (d) in a 2-dimensional surface topography and in (c) and (f) the corresponding 3-dimensional representation. The micrographs (b) and (e) show the magnetic structure (phase diagram) of the same regions. The micrographs (a)-(c) show one region of the sample with a hierarchically twinned microstructure on three levels similar to the schematic in Figure 3.20a. While the topography micrograph shows fine twins on different levels and clear twin boundaries, the magnetic structure is spanning over twin boundaries. The micrographs (d)-(f) show a twin structure with larger surface twins (d) than the recorded magnetic structure (e). The vertical magnetic structure visible in the top right half of (e), is only very faintly visible in the topography micrograph (d).

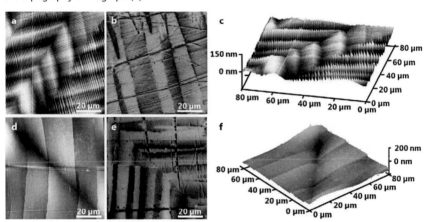

Figure 3.13: Twin microstructure of a $(Ni_{51}Mn_{28}Ga_{21})_{99.5}Dy_{0.5}$ sample [36,37]. The top views are given in (a) and (d), the corresponding magnetic structures in (b) and (e), and a three dimensional views in (c) and (f). Three different hierarchies of twins are present. There are four large twins with a thickness of 20 - 40 μm and twin boundaries extending from bottom left to top right. The large twins accommodate intermediate twins with thicknesses between 5 and 15 μm. The intermediate twins form a zigzag pattern across the boundaries of the large twins. Within the intermediate twins there are small twins with thicknesses of 1 to 2 μm. In (d) to (f), larger vertical twins of 20 μm, a diagonal boundary, and two horizontal twin boundaries are visible. Along the diagonal boundary, the direction of magnetic domains is changing from vertical (bottom left half) to horizontal (top right half). The angles between the twin slopes of the small twins indicate the 14M structure [60,61].

3.4.3. PHASES AND STRUCTURES OF THE NI-MN-GA SYSTEM

The structure of Ni_2MnGa depends on temperature and composition. Upon cooling, several phase transformation occur with increasing order towards lower temperatures [62], which can be seen in the phase diagram shown in Figure 3.14. The phases are given in the strukturbericht notion in Figure 3.14. The first phase transformation is from the liquid to a high temperature disordered A2 phase (body centered cubic structure) below 15 at.-%. This phase transformation was not observed for $Ni_{50}Mn_xGa_{50-x}$ (with $15 \leq x \leq 35$). In the A2 phase all elements occupy lattice point in random order. For the composition range with $15 \leq x \leq 35$, a phase transformation from liquid to B2' phase occurs, which has a simple cubic structure. The B2' (prime) phase is more ordered than the A2 phase, since the Ni atoms are occupying the body center, while the Ga and Mn atoms are sharing the corner sites at random. This results in a simple cubic unit cell. Depending on composition, the solidus temperature is between 1060K and 1730K [62,63]. The next phase transformation is from B2' to $L2_1$. The $L2_1$ phase has a cubic phase centered lattice with a $Fm\overline{3}m$ space group (225) (see [47,48]). When cooling further, the $L2_1$ phase is transforming into a martensite phase. The martensite phase may exhibit a lattice modulation with a stacking sequence depending on the composition. The most common stacking sequences are 14M ($\overline{5}2)_2$ and 10M ($\overline{3}2)_2$ [5]. More information about the stacking nomenclature can be found in [64]. As described in section 3.4.2.3, the ground state of the martensite phases (upon cooling and increasing applied stress) is a nonmodulate martensite phase (NM) with no stacking.

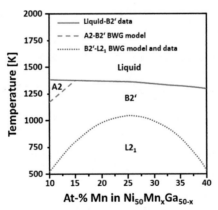

Figure 3.14: Experimental quasibinary temperature-composition diagram for $Ni_{50}Mn_xGa_{50-x}$ for the stoichiometry range $15 \leq x \leq 35$. Dashed and solid lines indicate theoretical predictions of phase transformations, dots represent experimental data. The red line marks the liquidus line. The prime in B2' indicates that Mn and Ga are sharing the corner sites. Graph based on data and graph from [62].

The space group of the stacked monoclinic martensite is when following the International Table of Crystallography, C2/m. Depending on the properties that one wants to emphasize, different unit cells can be chosen. The most common four unit cells for the **14M martensite** will be described here in more detail and shown in Figure 3.15.

When emphasizing the stacking of the 14M martensite, monoclinic unit cells are chosen to describe the martensite lattice. The **b** axis of the monoclinic unit cells is identical with the c_m axis of the fcc unit cell of the L2$_1$ phase. The **a** and **c** axes though are rotated by approximately 45° around the **b** axis, so that the stacking planes are parallel to the (001) planes. All monoclinic unit cells have a mirror plane in the **a** x **c** plane. The monoclinic unit cell that describes best the stacking sequence of the 14M martensite is here called 14M$_1$.

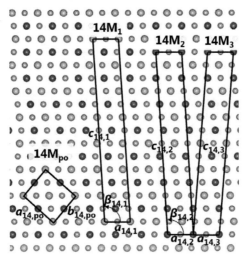

Figure 3.15: Four 14M unit cells: 14M$_{po}$, 14M$_1$, 14M$_2$, and 14M$_3$. The 14M$_{po}$ is based on the cubic unit cell of the L2$_1$ phase and averaged of the 14M lattice. The 14M$_1$ unit cell best describes the ($\overline{5}$ 2)$_2$ stacking, while the 14M$_3$ best represents the symmetry of the 14M lattice. The 14M$_2$ is an intermediate unit cell between the 14M$_1$ and 14M$_3$.

The bottom (001) plane of the 14M$_1$ unit cell the lowest of the five left shifting (001) planes with Ga or Mn atoms occupying the corner sites of the unit cell. From bottom to top of the unit cell, five (001) planes shift to the left, then two to the right, again five to the left and two to the right. In the 14M$_1$ unit cell, the corner site atoms also occupy the body center. When shifting the origin of the 14M$_1$ unit cell one (001) plane down, the 14M$_2$ unit cell is drawn. The 14M$_2$ unit cell has the following stacking of the (001) planes: one to the right, five to the left, two to the right, five to the left, and one to the right. The 14M$_2$ unit cell has an I2/m bravais lattice due an additional two-fold rotation axis parallel to the **b** axis through the body center (I). The $14M_2$ unit cell still follows like the $14M_1$ unit cell the stacking sequence along its **c** axis. When tilting the $14M_2$ unit cell to the right, this new $14M_3$ unit cell has the C2/m bravis lattice of the 14M lattice described in the International Tables of Crystallography and therefore describes best the symmetry of the 14M lattice. The $14M_3$ unit cell is side centered (C). Since shape change of the 14M lattice cannot be easily derived from the monoclinic unit cells, a forth unit cell based on the cubic axes system of the austenite $L2_1$ phase is derived. This pseudo-orthorhombic unit cell, here called $14M_{po}$, is averaged over all 14 stacking planes. All three monoclinic and the pseudo-orthorhombic 14M unit cells are shown in Figure 3.15 and lattice parameters of the pseudo-orthorhombic and $14M_1$ unit cells in Table 3.4, e.g. [65].

Figure 3.16 shows the $14M_1$ unit cell in more detail. The unit cell viewed along $[100]_{14M,1}$ is shown in (a). The unit cell viewed along $[010]_{14M,1}$ is shown in (b). The $(\bar{5}\,2)_2$ stacking sequence indicates the shift of $(001)_{14M,1}$ planes, five times to the left, two to the right and again five times to the left and two times to the right to complete the unit cell. In (c), the unit cell is viewed in a random direction.

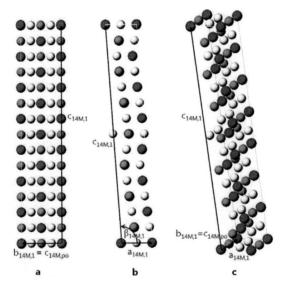

Figure 3.16: $14M_1$ unit cell with Ni (grey), Ga (blue), and Mn (green). The unit cell is viewed along $[100]_{14M,1}$ in (a). The unit cell is viewed along $[010]_{14M,1}$ in (b). The unit cell is viewed in a random direction in (c). Figure based on [22].

Similarly to the 14M monoclinic and pseudo-orthorhombic unit cells, three monoclinic unit cells and a pseudo-tetragonal unit cell can be chosen for the **10M martensite**. The same criteria for choosing the 14M unit cells are given for the $10M_1$, $10M_2$,

$10M_3$, and $10M_{pt}$ (pseudo-tetragonal) unit cells. The difference between the 14M and 10M martensite lattice is the stacking of the (001) planes. The 10M martensite has $(\bar{3}\,2)_2$ stacking sequence, i.e. three (001) planes shift to the left, then two to the right, again three to the left, and two to the right.

Figure 3.17 shows the $10M_1$ unit cell in more detail. . The unit cell viewed along $[100]_{10M,1}$ is shown in (a). The unit cell viewed along $[010]_{10M,1}$ is shown in (b). The $(\bar{3}\,2)_2$ stacking sequence indicates the shift of $(001)_{10M,1}$ planes, three times to the left, two to the right and again three times to the left and two times to the right to complete the unit cell. Because of the small deviation from orthogonality in the 10M martensite, the small angle is emphasized by the red lines that indicate the stacking section directions. In (c) the unit cell is viewed in a random direction.

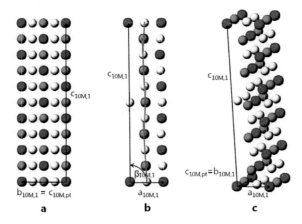

Figure 3.17: Crystal structure of the 10M martensitic Ni_2MnGa structure with Ni (grey), Ga (blue), and Mn (green). The unit cell is viewed along $[100]_{10M,1}$ in (a). The unit cell is viewed along $[010]_{10M,1}$ in (b). The red lines emphasis the angle of the stacking section. The unit cell is viewed in a random direction in (c). Figure based on [22].

The unit cell used to describe the **nonmodulated martensite (NM)** is based like the pseudo-orthorhombic and pseudo-tetragonal unit cells on the austenite $L2_1$ unit cell. The nonmodulated ground state of the martensite phase is tetragonal with c > a and a easy magnetization plane in the a_{NM} x b_{NM} plane and the hard magnetization direction along the c_{NM} axis.

In opposite to the NM martensite, the $c_{14,po}$ and $c_{10,pt}$ axes are the direction of easy magnetization. Later in this thesis, the ratio $c_{14,po}/a_{14,po}$ and $c_{10,pt}/a_{10,pt}$ - also called tetragonality - of the martensitic unit cell will be used since it describes the maxium possible MFIS of a martensite phase. The pseudo-orthorhombic and pseudo-tetragonal unit cells are in this dissertation used to index the fundamental reflection of the neutron hk0 planes. An overview of the bravais lattice and space group of the $L2_1$ phase and the martensite $14M_2$, $14M_3$, $10M_2$, and $10M_3$ unit cells are also given in Table 3.3. The unit cell parameters for the $14M_1$, $14M_{po}$, $10M_1$, and $10M_{pt}$ unit cells are also given in Table 3.4 [66].

Table 3.3: Classification of the three Ni-Mn-Ga structures relevant for this study. The first column gives the phase of Ni-Mn-Ga, the second the most common stacking sequences, the third the Bravais lattice of the unit cell, and the forth the space groups in three different writings.

Phase	Index	Bravais Lattice	Space Group	
Austenite	-	Cubic	$Fm\overline{3}m$	225
Martensite	$10M_2$	Monoclinic	$I2/m$	12
	$14M_2$	Monoclinic	$I2/m$	12
	$10M_3$	Monoclinic	$C2/m$	12
	$14M_3$	Monoclinic	$C2/m$	12

Table 3.4: Lattice parameters and unit cell angles of the $L2_1$, 10M, 14M, and NM Ni-Mn-Ga structures in the different unit cell representations. The first column gives the present phase, the second the stacking sequences, the third the lattice parameters, and the forth column the angles of the unit cells and c/a ratios [65,66].

Phase	Unit cell	Lattice Parameters [Å]			Angles [°]		
		a	b	c	α	β	γ
Austenite	-	5.82	5.82	5.82	90	90	90
Martensite	$10M_1$	4.24	5.66	20.5	90	90.5	90
	$14M_1$	4.23	5.51	29.4	90	93.5	90
	$10M_{pt}$	5.94	5.94	5.59	c/a = 0.94		
	$14M_{po}$	6.12	5.80	5.50	c/a = 0.90		
	NM	5.46	5.46	6.58	c/a = 1.21		

New approaches to describe the Ni-Mn-Ga martensite structure, e.g. by Kaufmann et al. [67], are directed towards an interpretation of the modulation as nanotwinned tetragonal or 2M lattice.

3.4.4. FERROMAGNETISM OF NI$_2$MNGA

The ferromagnetism of Ni_2MnGa is "largely confined to the Mn sites" [48]. The simulated and experimentally determined magnetic moments of three stoichiometric Heusler alloys are listed in Table 3.5 together with the lattice parameter, a, of the $L2_1$ phase calculated with the generalized gradient approximation method [50]. In the third column the total magnetic moment, μ_{tot}, (experimental values are in parenthesis) is given. The magnetic moment of each element, μ_i is given in the fourth column. It is also shown in [50] (see Table 3.5) that the magnetic moment of Ga is negligible, while Mn has a nearly ten times higher magnetic moment than Ni. This is interesting because in its elemental form, Mn is antiferromagnetic. The change from the antiferromagnetic behavior of Mn to ferromagnetic behavior is due to the increased distance between the Mn sites in the $L2_1$ structure compared to the distance between Mn atoms in pure Mn. The increased interatomic distance changes the Mn-Mn exchange interaction from antiferromagnetic to ferromagnetic [50].

Table 3.5: Calculated properties of Ni_2MnX Heusler alloys for X = Al, Ga, Sn [50]. The first column of this table shows the composition of the Heusler alloy, the second column its lattice parameter, a, the third column the total magnetic moment, μ_{tot}, (experimental values are in parenthesis), and the forth column the magnetic moment of each element, μ_i. Data in table based on [50].

Ni₂MnX	a [Å]	μ_{tot} [μ_B]	μ_i [μ_B]		
			Ni	Mn	X(=Al,Ga,Sn)
Ni₂MnAl	5.81 (5.84)	4.03 (4.19)	0.38	3.30	-0.06
Ni₂MnGa	5.83 (5.84)	4.09 (4.17)	0.37	3.36	-0.04
Ni₂MnSn	6.08 (6.07)	4.08 (4.05)	0.24	3.53	-0.03

Ni_2MnGa in its martensite phase has as described in section 3.3.6 also magnetic anisotropic properties. The easy magnetization direction is along [001] or the c direction of the pseudo-orthorhombic and pseudo-tetragonal martensite phases. The nonmodulated phase (NM) has an easy magnetization plane perpendicular to the [001] or c direction of the nonmodulated phase. Table 3.6 shows experimental values of the saturation magnetization, M_s, and the magnetic anisotropy between the c axes and the a and b axes (index of K) of the martensite phases [68].

Table 3.6: Lattice parameters and magnetic properties of the 10M, 14M, and nomodualted (NM) martensite phases. The magnetic anisotropy is given between the c axis and the a and b axes (index of K). Additional, the saturation magnetization is given [68].

phase	a [Å]	b [Å]	c [Å]	Magnetic anisotropy, K		M_s
				K_a [10^5 J/m³]	K_b [10^5 J/m³]	[10^5 A/m]
10M	5.94	5.94	5.59	1.45	1.45	4.8±0.4
14M	6.19	5.80	5.53	1.6	0.7	4.8±0.4
NM	5.46	5.46	6.58	-2.03	-2.03	4.8±0.4

3.5. MAGNETOPLASTICITY

Ni-Mn-Ga MSMA show very large magnetoplasticity. The magnetoplasticity is responsible for the magnetic shape memory effect. This section is divided in three paragraphs describing the magnetoplastic behavior from the macroscopic, mesoscopic, and microscopic perspectives. It is assumed that the MSMA sample is at a temperature below the martensitic transformation if not otherwise stated.

3.5.1. MACROSCOPIC APPROACH

A macroscopic approach of the magnetoplasticity of MSMA describes the phenomenology, which is a magnetic-field-induced strain (MFIS). When the magnetic field strength is increased or the direction of an external magnetic field is changed, the MSMA changes its macroscopic shape by a certain amount. While the sizes in one or two directions of the MSMA sample increase, the sizes in the remaining directions decrease until they reach saturation. When the magnetic field is removed, the macroscopic shape of the MSMA do not recover to the initial state. The magnetic-field-induced deformation is mostly plastic. The initial macroscopic shape can be restored by applying a compressive force parallel to or by applying a magnetic field perpendicular to the initial field.

In a rotating magnetic field, the shape change is periodical (see Figure 3.3). During one full rotation of the magnetic field the shape is changing two times from its initial dimensions to a state of maximum change and back to the initial shape. One 360° rotation of the magnetic field contains two magneto-mechanical cycles of each 180°. This phenomenon is caused by the alignment of the axis of easy magnetization, which is the c direction of the pseudo-tetragonal (10M) unit cell. Due to the tetragonal lattice and a ratio of $c/a < 1$, this causes a macroscopic change of length of the same ratio, if the entire crystal is completely aligned in both, the initial and the saturated state (see Figure 3.18).

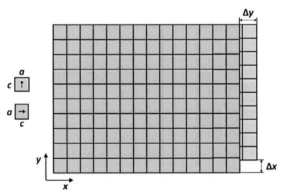

Figure 3.18: Schematic macroscopic shape change of a single crystal bulk MSMA. The MSMA is shown with two different saturated states: The blue "crystal" has its unit cells aligned with their c direction vertically. The orange "crystal" has its unit cells aligned with their c direction horizontally. The ratio of the macroscopic shape change is similar to the c/a ratio of the pseudo-orthorhombic and pseudo-tetragonal unit cells.

To reach the maximum shape change, the MSMA needs to be completely in a saturated or compressed state, so that all c directions are aligned. The MSMA consists then only out of one domain. This alignment can be performed by applying a magnetic field or stress parallel to the {100} type direction in the martensite phase or during the martensitic phase transformation. The maximum theoretic magnetic-field-induced strain is $\varepsilon_{max} = 1 - c / a$. The c/a ratio depends on the crystal structure of the martensite phase. For the 10M structure c/a is 0.94 and for the 14M structure 0.90.

3.5.2. MESOSCOPIC APPROACH

On a mesoscopic scale, the magnetoplasticity of MSMA takes place through the motion of twin boundaries.

Figure 3.19 illustrates the mechanism of twin boundary motion which is necessary for the alignment of the c direction of the unit cell parallel to the magnetic field or compressive stress.

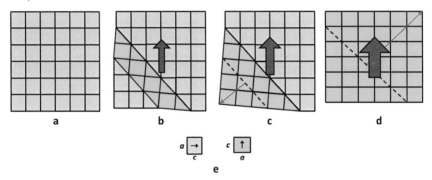

Figure 3.19: Schematic of twin boundary motion. The initial, single domain twin structure with its c axis aligned horizontally is shown in (a). When applying a magnetic field or mechanical stress vertically (indicated by the blue arrow), somewhere in the sample twin boundaries are moving and realigning the c axis and magnetic moment vertically (b) and (c) until another single domain state is reached (d) with all c axes aligned vertically. The orientation of the unit cells and magnetic moments (black arrow) depending on their color are indicated in (e). Twin boundary motion is indicated by the red arrow.

The initial state of a single twin variant state with no applied magnetic field or stress is shown in (a). When a magnetic field or stress is applied (blue arrow in b), the magnetic moments tend to align with the field. To align the magnetic moment of the unit cells to the magnetic field, the unit cell has to be reorientated with its c direction parallel to the magnetic field or stress. This happens with the movement of the twin boundary.

Twin boundaries can move within the entire sample simultaneously if the magnetic or mechanical force is higher than the repulsive elastic force (see paragraph 3.5.3). A twin boundary can move through an entire martensite variant and realign all magnetic moments parallel to the magnetic field until stopped at grain boundaries or defects.

3.5.3. MICROSCOPIC APPROACH

On a microscopic scale, twin boundary motion is caused by twinning dislocation motion. The twin boundary movement is initiated at one location of a coherent twin boundary. A schematic of twin variants (with variant boundaries α and β) and twins (A, B, C) within the twin variants is shown in Figure 3.20 [69]. In the detail view, the orientation of the easy magnetization direction in each twin is marked with c_A (horizontal) and c_B (vertical). With an applied force or magnetic field, a favorable twin consumes a less favorable twin. If in Figure 3.20 a magnetic field or mechanical force is applied in x direction, twin A with c_A parallel to x is favorable, because c_A is the direction of easy magnetization and also the shortest lattice parameter. The twin boundary moves by the motion of a step at the coherent twin boundary, which is called a disconnection. A disconnection is characterized by a Burgers vector, b_t (similar to a dislocation), and a step heights, d_t, [70]. When disconnections move along a twin boundary, one twin is growing on expense of the other. Moving the disconnection requires a force, which can be induced by a magnetic field or stress. An applied magnetic field is causing a magnetic force, F_M, on these disconnections. With M being the saturation magnetization of the twin variant and H the magnetic field, θ_A and θ_B being the deviation between M and the axis of easy magnetization, γ, the angle between the x axis (parallel to c_A, Figure 3.20) and the magnetic field, and K_x (x = 1,2) being the magnetic anisotropy constant, the magnetic density u_A and u_B of both twins A and B can be calculated with [71]:

$$u_A = -\mu_0 MH \cos\left(\gamma - \theta_A\right) + K_1 \sin^2 \theta_A \tag{3.7}$$

$$u_B = -\mu_0 MH \cos\left(\gamma - \theta_B\right) + K_2 \sin^2 \theta_B \tag{3.8}$$

With $H \gg H_A$, the angle θ_A and θ_B can be approximated with γ, and the energy created by the magnetic field is:

$$u_A = K_1 \sin^2 \gamma - \mu_0 MH \tag{3.9}$$

$$u_B = K_2 \sin^2 \gamma - \mu_0 MH \tag{3.10}$$

If the magnetic field is parallel to the easy magnetization direction of twin A, c_A, γ become zero and the energy difference Δu between twin A and B is:

$$\Delta u = u_A - u_B = -K_1 \tag{3.11}$$

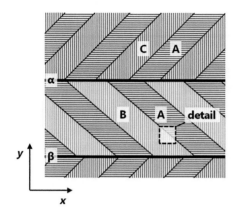
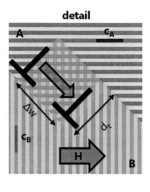

Figure 3.20: Schematic of twin variants (with variant boundaries α and β) and twins (A, B, and A, C) within the twin variants separated by the variant boundaries. Detail illustrates the orientation of the easy axis of magnetization, c, in twin A and B. The detail shows a schematic of twin interface in a magnetic field with disconnection. d_t is the step height and dx the disconnection moving distance. Based on [69].

This means, that the energy of twin A is lower than in B and therefore twin A is growing by consuming twin B. For $\gamma = \pi/2$ or $3\pi/2$ twin B grows on cost of twin A. For a full rotation of the magnetic field, twin A grows and shrinks twice. This is the reason why a magneto-mechanical cycle is complete with the rotation of the magnetic field by 180° (Figure 3.3). The force, F_M, that is produced by a magnetic field, which is parallel to the x or y axis on the twinning disconnection is [35]:

$$F_M = d_t K_1 \tag{3.12}$$

The magneto-stress, τ_M, on the twinning plane in twinning direction is a shear stress. With s being the twinning shear $s = d_t / b_t$, the maximum magnetostress is:

$$\tau_M = \frac{K}{s} \tag{3.13}$$

The magnetic field exerts a magnetic force on the disconnection, which causes the disconnection to move. The motion of the disconnections finally leads to the motion of the twin boundary and the growth of one twin and shrinking of the other. The motion of the disconnection and twin boundary results in a small shear (Figure 3.21).

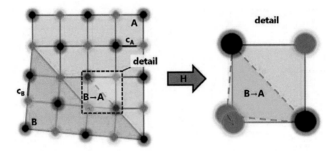

Figure 3.21: Shear during disconnection motion. In a horizontally applied magnetic field or stress, twin B gets consumed by twin A through twin boundary motion which results in a small shear of the atoms of twin B along the diagonal twin boundary.

4. EXPERIMENTAL METHODS

4.1. SPECIMEN FABRICATION

The preparation of single crystalline Ni-Mn-Ga samples and ingots the samples are cut from are important steps to create samples that perform. The repeatability of certain properties, most importantly control of composition that determines most all other properties of the Ni-Mn-Ga MSMA, is thereby essential. The following sections describe all steps necessary for the production of the samples.

4.1.1. GROWING OF SINGLE CRYSTALS

The single crystals were grown, using the patented slag remelting and encapsulating (SLARE) method [72]. The SLARE method is based on the Bridgman single crystal growth method [73]. Figure 4.1 shows a schematic of a (a) Bridgman and a (b) SLARE method. During the growth of single crystals using the Bridgman method, a polycrystalline rod is placed in a crucible with a cone shape tip at the bottom. The crucible is positioned in a vertical tube furnace heated 50 to 100 K above the melting temperature of the, to be melted, material. By slowly lowering the crucible with the molten metal, the temperature decreases at the bottom of the crucible. This eventually leads to the crystallization of a few seed crystals (in the best case only one) in the tip of the crucible. By further lowering the crucible out of the heated zone of the Bridgman furnace, the mold is slowly solidifying with the orientation of seed crystal. If only one seed crystal is created, a large single crystal is grown. Due to a flattening temperature gradient towards the center axis of the crucible, defects are more likely to appear in the center, especially when the crucible diameter is increased. The growth process is normally taken place under a noble gas atmosphere to prevent reactions of the liquid material and after crystallization with the heated solid material.

There are two crucial differences between the SLARE method and Bridgman method: first, a slag is surrounding the molt to build a liquid agent. Second, the solidification of the molt is realized by the decrease of the furnace temperature instead of the motion of the crucible. The slag has to be chemically inert against the molten metal alloy and the crucible. Impurities, e.g. oxides and sulphides, should be on the other hand highly soluble in the slag, so that the impurities move from the metal to the slag and the metal becomes purified. Additionally, the liquidus temperature of the slag has to be below the solidus temperature of the alloy. To cover the molten alloy, the density of the slag has to be lower than the density of the alloys. Another advantage of the slag lies in the capability to prevent the inert gas to build bubbles within the metal alloys. It also prevents the Mn, which has a low vapor pressure, from evaporating during the single crystal growth process. The loss of Mn is one of the challenges during the single crystal production of Ni-Mn-Ga ingots.

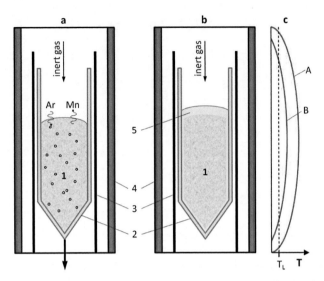

Figure 4.1: Schematic of (a) Bridgman method and (b) SLARE method. In both methods, a metal alloys ingot (1) is placed in the crucible (2) positioned in a ceramic tube (3) that is flushed by inert gas (e.g. argon) within a vertical tube furnace (4). Using the SLARE technique (b), a slag (5) is purifying the metal alloys while preventing Mn loss and gas bubbles. During the initial heating step A in (c), the furnace temperature is raised above the liquidus temperature (T_L) of the metal alloy throughout the entire crucible. The following temperature decrease is realized in (a) the Bridgman method by slowly lowering the crucible and in (b), the SLARE method by slowly decreasing the furnace temperature (B).

In the SLARE technique furnace, a polycrystalline ingot was placed into a crucible packed with a slag, which was matched with the crucible and ingot material. The furnace was then heated in vacuum to 300 °C and kept at this temperature for 90 min. This step was chosen so that moisture on the polycrystalline ingot evaporates. The furnace was then heated to 1,250 °C, where it was held for 20 min in a 5 mbar H_2 atmosphere. The atmosphere was then changed to a 10/90 H_2/Ar mixture. The crucible tip temperature was then lowered to 1,180 °C for 30 min, the furnace evacuated and flushed again with the 10/90 H_2/Ar mixture. The crucible tip temperature was then further lowered with 3 K/h to 840 °C.

A schematic of the temperature profile of the high temperature state (A) and the crystallization state (B) is shown in Figure 4.1b. (A) shows that the temperature is over the entire length of the metal alloy ingot above the liquidus temperature of the Ni-Mn-Ga alloy. After the entire Ni-Mn-Ga ingot was melted, the temperature of the furnace was lowered, so that the temperature in the tip of the crucible is below the crystallization point of Ni-Mn-Ga.

4.1.2. ANNEALING

To reduce the composition gradient in the grown single crystal ingot, the ingot is sealed in an Argon filled quartz tube under Argon atmosphere and annealed. The sample was heated in a furnace with 300 K/h to 1000 °C and held at 1000 °C for 80 h. In a second step, the temperature was decreased with a rate of 300 K/h to 725 °C and held at 725 °C for 2 h. For the third step, the oven was cooled with a rate of 10 K/h to 700 °C and held at 700 °C for 10 h. During the fourth step, the oven was cooled further to 500 °C with a rate of 50 K/h and held at 500 °C for 20 hours. Finally, the oven was cooled to room temperature with a rate of 50 K/h.

4.1.3. CRYSTAL ORIENTATION AND CUTTING

All samples presented in this work were oriented with their surfaces parallel to the $\{100\}_{austenite}$ planes. All "Berlin samples" were spark eroded as described in section 4.1.3.1. All samples were received already in an oriented state and most of them (all "Berlin samples") already cut in parallelepipeds of approximately 6 mm x 4 mm x 3 mm. The Berlin single crystals were oriented with a Laue Camera and cut with wire electrical discharge machining (see section 4.1.3.1) at room temperature with the lowest power setting by MaTecK[2]. The following sections describe the two sample cutting methods used in this study in more detail.

4.1.3.1. SPARK EROSION CUTTING

Spark eroding or wire Electrical Discharge Machining (EDM) is a cutting technique using electro-thermic processes to remove electrical conductive material by spark discharges. Typically, during wire EDM the material are cut by a thin wire and submerged in a dielectric fluid (e.g. deionized water). During this process a high voltage impulse is applied between the wire and workpiece, which creates spark discharges. These discharges result in the removal of material along the wire. The constant translation of the wire leads to a uniform wear of the wire. Since the discharges are built between the wire surface and the workpiece, the cutting velocity is proportional to the wire diameter and also to the applied voltage. While the cutting velocity increases with increasing voltage and wire diameter, the surface quality is decreasing and the surface roughness increases. By shortening the voltage pulses, the surface quality is increasing. All electrically conducting materials can be cut or machined using EDM. The wires used for wire EDM have diameters between 10 μm and 0.25 mm.

The samples of this work were cut by MaTecK using wire EDM. The average surface roughness, R_a, of a spark eroded surface is approximately 1.6 μm (see section 7.3).

---

[2] Email conversation with H. Schlick, MaTecK- Material-Technologie & Kristalle GmbH in Jülich, Germany, 2nd January 2010

4.1.3.2. PRECISION WIRE SAW CUTTING

Because samples are produced in single crystal ingots and mostly received as slices or ingots and not in the necessary shape and size for magneto-mechanical testing, samples have to be cut to size and in case of the single crystal ingots in certain crystal orientations. The orientations were marked when the crystals were received.

It is important that the loss of material is minimized during the cutting process. The cut must be very precise. It is also important that the force on the sample during the cutting is as small as possible to receive a good surface quality. All these requirements are met by the K.D. Unipress Wire Saw type WS-22 with an attached goniometer. The sample can be mounted on the three-circle goniometer. The wire saw guides a 50 µm diameter Tungsten wire over the sample while an abrasive 3:1 oil-SiC slurry is dropping at a constant rate on the sample. The cut with the wire saw results in a homogenous surface finish as shown in the scanning electron microscope micrograph in Figure 4.2. The average surface roughness, R_a, of a surface cut with the precision wire saw surface is approximately 0.4 µm (see section 7.3).

Figure 4.2: The scanning electron micrograph shows the surface of a Ni-Mn-Ga sample cut with a precision wire saw Unipress type WS-22. The surface is very homogeneous and has no grooves.

4.1.4. SURFACE PREPARATION

Electropolishing was one method used to smoothen the surface of samples. Two different types of electropolishing were utilized to remove the damaged surface of the spark eroded samples. Except for the Berlin005 RP specimen, all samples were electropolished at the HZB with a standardized **procedure A** developed by MSM Krystall GbR. This procedure is divided in two parts: initial cleaning and electropolishing. The steps of the initial cleaning after the samples have been spark eroded are given in Table 4.1. After the initial cleaning, the samples were glued to a 2 mm diameter and 100 mm long copper rod with cyanacrylate based instant glue. Then conductive silver is placed around the copper-sample interface. After drying, the interface and the bottom 40 mm of the copper rod were coated with Lacomit G371 to isolate the conductive silver and copper rod from the electropolishing solution.. The electrolyte consisted of 3/4th of ethanol and 1/4th of 65% HNO_3, was cooled to -5 °C and constantly stirred The samples were dipped in the electrolyte two sequences, the first time for 90 s and the second time for 30 s. After each polishing sequence, the samples were dipped in 5% NH_3, washed with distilled water, then 5% HCl, and washed again with distilled water. Finally, the coating and conductive silver was removed with acetone. During the electropolishing treatment, a 1 mm thick, 100 mm x 20 mm aluminum sheet was used as cathode (-) submerged in the electropolishing solution while the sample was attached to the anode (+). The current density was chosen to be 0.006 to 0.01 A/mm^2, preferably 0.008 A/mm^2.

Table 4.1: Electropolishing procedure A cleaning steps.

Time	Ultrasonic bath mixture	Rinse with	Drying
20 min	5% Tickopur RW 77 in dist. H_2O	dist. H_2O	not performed
5 min	dist. H_2O	dist. H_2O	not performed
5 min	Acetone	Acetone	not performed
3 min	Ethanol	Ethanol	cotton wipes

Procedure B was used at BSU to electropolish samples Berlin005 RP (see section 4.4.1). During procedure B, samples were electropolished in three sequences, during the first two sequences for 60 s and during the third sequence for 120 s using a 60% sulfuric acid with a current of approximately 0.035 A/mm^2 and a voltage of 7 V.

Besides electropolishing, **mechanical polishing** was used to smoothen surfaces. To create a surface with a very low surface roughness, all samples were ground if bigger cracks were visible and then polished. For this procedure, a Struers LaboPol 5 was used. In case of visible surface cracks or inhomogeneous artifacts, the surface was ground with a grit 1200 SiC grinding paper. After 15 s of grinding or polishing, the surface was examined with an optical microscope. If no cracks or groves from prior grinding or polishing steps were visible, the next finer polishing step was performed where the sample was turned by 90°. This insures that groves created by the prior polishing step, are polished away. The polishing steps are given in Table 4.2. Each of the steps was repeated until the groves of the prior polishing step were completely removed.

Table 4.2: Mechanical polishing steps.

Step	Polishing Cloth	Polishing Slurry
1	Cameo Disk, Cameo Silver	Diamond liquid, 6Mme, 6 µm diamond suspension
2	Struers, MD Dac	Struers, DiaPro Dac, 3 µm diamond suspension
3	Struers, MD Nap	Struers, DiaPro Nap R, 1 µm diamond suspension
4	Struers, MD Chem	Struers, AP-A Suspension, 0.3 µm alumina suspension
5	Struers, MD Chem	Struers, OP-AA Suspension, acidic alumina suspension

4.2. CHEMICAL, MAGNETIC, AND THERMAL CHARACTERIZATION

4.2.1. OPTICAL MICROSCOPY

Two optical light microscopes were used to image the surface structure of the material. A Zeiss Axiovert 200 MAT with a Zeiss AxioCam attached, provides the option to view details of a sample's surface on a PC screen and to record pictures digitally. A disadvantage of this microscope is that with the attached oculars the lowest magnification is not low enough to get an overview micrograph of the entire sample. Overview picture were taken optical camera with 21 MPixel resolution (see section 4.4.3). Additionally, micrographs of the Zeiss optical microscope were merged to one large micrograph.

4.2.2. SCANNING ELECTRON MICROSCOPY AND ENERGY DISPERSIVE X-RAY SPECTROMETRY

For detailed crack and surface analysis in the micron and submicron range the scanning electron microscope (SEM) LEO 1430VP (Zeiss) was used. Additionally to the ability of detecting secondary electrons, the SEM is also equipped with an Oxford Energy Dispersive X-ray Spectrometer (EDS) and an Electron Backscattered Detector. The LEO 1430VP is using a Tungsten hairpin type filament. The following paragraphs are based on refs. [74,75].

The **secondary electrons** are created at the surface of the sample by an inelastic interaction of the incoming electron beam with the sample. For a secondary electron (SE) to escape the sample it is necessary to have an energy higher than the binding energy, normally a few eV. The Fermi energy level is the highest energy level of electrons in a metal at a temperature of 0 K. One primary electron is able to release at least one secondary electron, due to the low energy of SE. Most of the time, a primary electron can release more than one SE. The low energy of SE makes it also easy to collect them with a low bias voltage between sample and SE detector. Nearly 100% of the produced SE can be collected. Another effect of the low energy of SE is that they are not able to escape through a thick layer of solid material. This means that SEs are only escaping from the surface and from a depth of 1-2 nm. The signal of the SE is mostly influenced by the number of incoming electrons and the surface topography especially the angle between the surface, detector and incoming beam. On the other hand, it is very little influenced by the sample's composition or density. The resolution of current SE microscopes can be as good as 2 nm, compared to older SE microscopes with 20 nm resolution [74].

Energy Dispersive X-ray Spectrometry (EDS or EDX) characterizes x-rays that are generated during the interaction of primary electrons with the material they penetrate. When a primary electron ejects an electron of an inner shell, the vacated energy level of that electron will be filled by electrons from outer shells with higher energy. While relaxing from a higher energy level, energy is released in form of an x-ray. Because of the specific energy levels of the electrons in each material, the x-rays are specific to the element. With EDS, it is therefore possible to qualify and quantify different elements in a sample. EDS is limited by the energy resolution of the detector. Another limitation is given in the detection of light elements. The current detection limit for windowless detectors is Boron (Z = 5) [75]. Because x-rays can penetrate solids better than electrons, x-rays created by scattered electron in the entire interaction volume will be detected. This not only increases the number of detectable x-rays but also the volume the x-rays will be emitted by a very pointed incoming beam which decreases the spatial resolution of the elemental analysis. The size of the interaction volume depends on the energy of the incoming electron beam, the thickness of the sample, and the atomic number of the specimen.

The relative difference of chemical composition in over a series of samples can be detected with high accuracy if the same EDS conditions (i.e. working distance, acceleration voltage, spot size, beam current, sample surface polish) are used. The absolute chemical composition on the other hand can only be detected with a relative error of ±6%, if the element concentrations are more than 10 at.-%. The relative error of ±6% can only be achieved if the EDS system was previously calibrated using a sample with a known composition close to the one examined. This relative error is based on the computational correction of three factors that are influencing the concentration values of a sample: atomic number, absorption, and fluorescence.

To calibrate the EDS, a Ni-Mn-Ga sample was cut in two slices. One slice was again cut in four pieces. These pieces were analyzed by the SGS Institut Fresenius GmbH, Germany, using inductive coupled plasma optical (sometimes also: atomic) emission spectroscopy (ICP-OES or ICP-AES). The second slice was used to calibrate the EDS. The results of the ICP-OES analysis are given in Table 4.3. There was no systematical error included in the measurement protocol. The relative error for all composition ranges detectable by ICP-OES (detection limit is ppm) is 1% [76]. The systematical error and confidence region of the analysis results in an absolute error of the calibration sample of ±1.3 at.-%.

Table 4.3: Analysis protocol of reference sample.

Reference sample piece	Composition [at.-%]		
	Ni	Mn	Ga
1	49.4	26.8	22.8
2	50.3	27.3	23.8
3	49.3	26.6	23.6
4	50.2	27.5	23.4
average	49.8	27.1	23.6
standard deviation	±0.5	±0.4	±0.2
95% confidence region	±0.8	±0.7	±0.3

With this large absolute error in mind, two polycrystalline samples with nominal compositions of $Ni_{52.0}Mn_{24.4}Ga_{23.6}$ and $Ni_{46.7}Mn_{34.0}Ga_{19.3}$ were cast at BSU with a Reitel "Induret Compact" casting furnace. A weight loss of 0.7 at.-% during casting was attributed to Mn loss, due to the low vapor pressure of Mn. With this, the compositions of the both samples were corrected to $Ni_{52.5}Mn_{23.7}Ga_{23.8}$ and $Ni_{47.2}Mn_{33.3}Ga_{19.5}$. Both polycrystalline samples were analyzed using the same conditions on the same EDS. A summary of the results including the average concentrations (out of eight measured spots) and standard deviation is given in Table 4.4. Except for the Mn concentration of sample 2, the corrected composition is always within the 95% confidence region of the average composition. This indicates that the relative error of the absolute composition is approximately 3% (largest confidence region rounded up). This is below the relative error of 6% of EDS given in refs. [74,75].

Table 4.4: Composition of EDS reference control measurements

	Sample 1 [at.-%]			Sample 2 [at.-%]		
	Ni	Mn	Ga	Ni	Mn	Ga
nominal composition	52.0	24.4	23.6	46.7	34.0	19.3
corrected composition	52.5	23.7	23.8	47.2	33.3	19.5
average	52.4	23.5	24.2	47.2	32.8	19.9
standard deviation	±0.2	±0.3	±0.4	±0.3	±0.2	±0.3
95% confidence region	±0.3	±0.4	±0.5	±0.4	±0.3	±0.4

From each single crystal (Berlin004, Berlin005, and Berlin054) at least one sample per slice was analyzed as well as all Berlin005RP samples. All measurements were performed with EDS of the Philips XL30 ESEM (HZB) with a working distance of 10 mm, spot size 4, 20 kV acceleration voltage, 200 counts per second and measurement time of 100 live seconds.

4.2.3. VIBRATING SAMPLE MAGNETOMETRY

Vibrating sample magnetometers (VSM) are used to determine the magnetization of a sample. Depending on the VSM, different conditions of the sample environment can be changed, e.g. magnetic field strength, magnetic field orientation, mechanical constraints, and temperature. To measure the magnetization, a magnetic sample is vibrated between pick-up coils, in which the sample is inducing a voltage, U_{ind}, according to Faraday's Law:

$$U_{ind} \sim -N\frac{d\Phi_M}{dt} \tag{4.1}$$

where N is the number of coil turns, Φ_M the magnetic flux, and t the time. Based on Faraday's law, the induced voltage is proportional to the rate of change of the magnetic flux through the coils. The changing magnetic flux is caused by the vibrating magnetic sample, while a static magnetic field which magnetizes the sample does not induce a voltage in the coils.

A VSM consists of several essential components as shown in Figure 4.3: the electromagnets (1), a hall probe (7) measuring the static magnetic field, pick-up coils (2) to measure the magnetization of the sample (5) that is connected through a glass rod (4) to a vibration unit (3). The glass bottle (6) is used to flush heated or cooled air towards the sample.

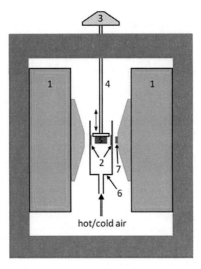

Figure 4.3: Schematics of a VSM. The electromagnets (1) produces a static magnetic field. The sample (5) is connected through a glass rod (4) to the vibration unit (5). The sample (5) is positioned within a glass bottle (6) that is flushed by heated or cooled air. At the height of the sample, pick-up coils are located around the sample within the glass bottle. A hall probe (7) is measuring the static magnetic field.

The primary function of a VSM is to measure the magnetization of a sample. By performing magnetization measurements at constant temperature and with constant field direction but with varying magnetic field, the magnetic hysteresis can be measured and M_{sat}, M_r, and H_a identified. By measuring the magnetic hysteresis in different direction, the magnetic anisotropy, K, can be determined. With a constant magnetic field and constant field direction but changing temperature, the martensite phase transformation can be characterized by identifying A_s, A_f, M_s, and M_f. Additionally, the Curie temperature, T_c, can be measured.

In this dissertation, a "Model 10 Vibrating Sample Magnetometer" was used (Figure 4.4) with a 2 T electromagnet (model 3474-140 by GMW magnet systems). Coupled with the current reverse switch, also the negative range of 0 to -2 T can be reached without rotating the electromagnet. The magnet can be rotated by 720° around the sample and glass bottle. The pole pieces can be adjusted so that the air gap between the magnets varies from 0 to

160 mm. During the measurements, the air gap between the pole pieces was kept constant at 40 mm. The vibration unit consists of two speaker-like units: one to vibrate the sample at 75 Hz, the other to insulate the sample from outside vibrations.

Figure 4.4: Photograph of the Model 10 Vibrating Sample Magnetometer. The left photograph shows the VSM itself and its controlling unit. The inset shows details of the VSM with the vibration unit (1), the windings of the electromagnet (2), the pole pieces (3), and the glass bottle (4) connected to the heated/cooled air with the sample on a glass rod in its center.

Martensitic Transformation and Curie Temperature: The heating/cooling unit consists of five parts: a controller unit with two thermocouples, a heating coil, and a cooling unit. Heating was realized by air flowing through the glass bottle from the bottom to the top. The air was heated within the glass bottle by a heating coil to up to 180 °C. The cooling of the sample was realized by cooled air, which flew through a liquid nitrogen cooled copper coil. The temperature was controlled by varying the heating current for the heating coils as well as the air flow through the liquid nitrogen cooled copper coil. A thermo couple in the air flow below the sample was used to read the temperature. The temperature of the sample depends greatly on the sample size, volume, porosity, and heat conductivity. Therefore, an additional thermocouple was mounted between the glass sample holder and the sample to monitor the sample temperature as shown in Figure 4.5. The deviation between the sample temperature and the sample environment temperature also depends on the heating and cooling rates. The air flow temperature is very sensitive on the thermocouple position within the glass bottle and needs to be calibrated every time the glass bottle is moved. Exemplarily, Figure 4.6 shows the temperatures at the sample and in the air flow. During heating the thermocouple in the air flow measured temperature 8 °C above and during cooling 5 °C below the sample temperature. This deviation appears as a ficticious hysteresis in the magnetization vs. temperature curve (Figure 4.6). When using the sample temperature, the magnetization vs. temperature curve has no hysteresis between the martensitic transformation and Curie temperature, as it should be.

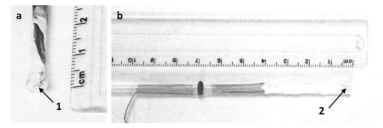

Figure 4.5: VSM thermocouple mount at the sample. (a) shows the thermocouple (1) without sample positioned at the glass sample holder. In (b) the sample (2) is mounted on top of the thermocouple and wrapped in Teflon tape.

Overall, the accuracy of the temperature measurement at the sample is assumed to be around ±1 °C. The Curie and phase transformation temperature measurements on bulk single crystals were measured in the following way: a relatively small magnetic field of 25 mT was applied at room temperature. The sample was then heated up to 140 °C at a rate of 5 K/min. After reaching 140 °C, the temperature was hold for 2 min and then reduced to room temperature at a rate of 5 K/min, where it was again hold for 2 min.

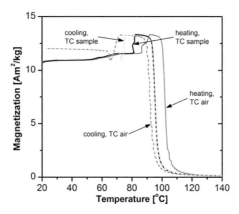

Figure 4.6: Comparison of thermocouple temperatures in VSM exemplarily shown at sample Berlin005 01C. Solid lines indicate temperatures during heating, dashed during cooling. Upon heating, the thermocouple in the air stream shows a temperature approximately 8 °C above the sample temperature, upon cooling 5 °C below. The temperatures measured at the sample during demagnetization at the Curie temperature are nearly identical, indicating an accurate temperature measurement.

To identify the **Saturation Magnetization**, the magnetization of a bulk single crystal was measured by increasing the magnetic field to 1.4 T (or 1.6 T for nonmodulated samples), well above the saturation field. The magnetic field was applied parallel to one set of pick-up coils. At this saturated condition, the sample was centered to eliminate errors from bending of the glass rod the sample is mounted to or the plastic connector that connects the glass rod (4 in Figure 4.3). The saturation magnetization is used to normalize the anisotropy measurements described below. Due to the importance of these measurements, the saturation magnetization was confirmed for two samples at a second VSM (Lakeshore) operated by a different user. The deviation was found to be smaller than 2%. The saturation magnetization measurements were always performed without additional sample holder attachments, e.g. magnetic anisotropy sample holder.

To determine the **Magnetic Anisotropy** a bulk single crystal sample was mechanically compressed in parallel to the longest sample dimension (z direction) before the measurement, so that a single variant state was created. This single variant state had the lattice parameter $c\|z$ and $a\|x\&y$ for a 10M martensite, and $c\|z$, $a\|x$ or y and $b\|y$ or x for a 14M martensite. The sample was mounted in a small constraining sample holder shown in Figure 4.7, which hindered the sample from deforming when a magnetic field was applied.

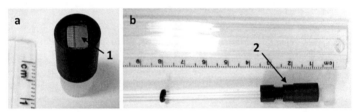

Figure 4.7: Custom made VSM sample holder for anisotropy measurements.

Initially, a sweep of 180° at 100 mT was used to identify the easy and hard magnetization direction, which are $<100>_{14M,po/10M,pt}$ type directions and parallel to the sample edges (c is the easy, a the hard). The magnetic field direction was then aligned to each $<100>$ direction (easy and hard for 10M and 14M martensite; for samples Berlin004 007C and Berlin005 05C (14M) also after remounting in the intermediate direction). The magnetic field was increased to 1.4 T for each direction and the magnetization curve were recorded. For the evaluation of the anisotropy energy, the curves had to be corrected with the demagnetization factor (see section 3.3.1). After the measurements were performed, the saturation magnetization at 1.4 T as described above was measured. The magnetization curves were finally normalized with the saturation magnetization measured separately without constraining sample holder. This was necessary because the sample was not perfectly centered in the compression sample holder and therefore not perfectly centered between the pick-up coils.

4.2.4. DIFFERENTIAL SCANNING CALOMETRY

During Differential Scanning Calometry (DSC) experiments two crucibles next to each other, one empty and one with a specimen are heated with the same temperature profile. The crucibles are placed in pans. At the bottom of each pan is an area thermocouple [77,78]. Due to the heat capacity of the specimen as well as phase transformations, the temperature of the crucible with specimen differs from the empty reference crucible. With a DSC, the heat flux in and out of a specimen at a specific temperature can be determined by comparing the reference and specimen temperatures.

The DSC used for the measurements shown here is a NETZSCH DSC 404 C, Erich NETZSCH GmbH & Co. Holding KG, Selb, Germany. The samples were heated under a constant flow of Argon from room temperature to 160 °C with 10 K/min hold for 3 mins and then cooled to 25 °C with 10 K/min. This initial heating and cooling is meant to relax the sample from constraints and to produce a selfaccommodated state of the sample. A second run with the same temperatures was performed with a heating and cooling rate of 5 K/min. Additional temperature profiles were run with nine samples (Berlin005 04C, 01D, 03D, 05E, 07E, 09E, RP11; Berlin047 05B, 01C) at heating and cooling rates of 2 K/min and 1K/min, to determine the influence of the heating rate on transformation temperatures.

The DSC measurements were used to systematically study the martensite transformation temperatures and Curie temperatures of all samples of single crystal ingots Berlin004, 005, 047, and 054. The martensite transformation temperature was determined by taking the average of the temperatures of the heating flux peaks during heating and cooling. The Curie temperature was determined by averaging the temperatures of the Curie peaks of the first derivative of the heat flux during heating and cooling. An example of a DSC measurement is shown in Figure 3.10.

4.2.5. X-RAY TOMOGRAPHY

A micro computer tomography instrument can be used to detect pores and cracks within a metallic sample. Higher density regions absorb more x-rays than lower density or hollow regions in the sample. An x-ray source emits x-rays towards the sample as illustrated in Figure 4.8. A flat panel detector is place behind the sample, and one picture is taken for each orientation of the sample, while the sample is rotating around 360°. From all images a 3-dimensional representation of the sample is reconstructed with the help of appropriate software. Various cross sections in any direction can be created and analyzed using graphics software. The micro computer tomographe at the HZB is equipped with a Hamamatsu flat panel detector and a micro focus tube with 100 kV acceleration voltage and a spot size of 7 µm. A 200 µm beryllium filter and a 1 mm aluminum filter were used to define the energy-range of the x-ray spectrum.

Figure 4.8: Schematics of the micro computer tomography setup. The x-ray source is shooting x-rays towards a rotating sample with 100 kV. The detector is placed behind the sample to collect the x-rays interacted with the sample.

Figure 4.9 shows the micro computer. The red x-ray source (1) is on the left, the sample (3) is place on a sample holder (2), and on the right side is the flat panel detector (4). During the reconstruction process, it is possible to reduce artifacts produced during the scan, e.g. effects of beam hardening (continuous change of shade of grey due to different energy absorptions caused by varying penetration depth) [79].

Figure 4.9: Photograph of the micro computer tomography instrument (a) and a detailed view of the sample holder (b). The x-ray source (1) is on the left, the sample holder (2) with the sample (3) is rotating while the flat panel detector (4) is collecting the intensity of the x-rays.

4.2.6. OPTICAL PROFILOMETRY

Optical profilometry was used to determine the surface roughness of samples at different stages of surface treatments. Optical profilometry is a profiling and surface characterization method using interferometry [80]. Thereby, a light beam from a single source is split in two beams with a semi-transparent mirror (Figure 4.10). While one beam is directed onto the surface, the other beam is relected of a reference mirror. Both parts of the beam are then recombined again constructive interference results in white and destructive interference black fringes. These fringes build topography like maps of the surface but do not reveal surface topography within them. Changing the focus slightly, the fringes are moving and surface feature that have been within a white or black fringe eventually appear at the fringes' interface. Using a CCD camera and data acquisition system, the moving fringes can be recorded and a topography map of the observed sample surface can be obtained.

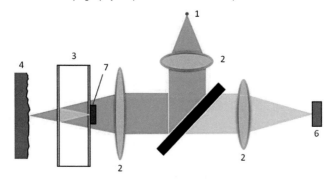

Figure 4.10: Schematic of optical profilometer. The light of a halogen lamp (1) get focused and directed (simplified optics (2) and beamsplitter (3)) towards a reference mirror within a Mirau objective (4) and the sample (5). The light from the sample and reference mirror is then focused on the CCD sensor (6). A split beam is reflecting from a reference mirror (7).

Two widely used operation techniques have been developed to fit specific applications: phase shifting interferometry (PSI) [81] and vertical scanning interferometry (VSI) [82]. PSI is ideal for smooth surfaces with a resolution in the sub-nanometer range. For this method, continuous fringes are evaluated to detect same height lines. Because of the monochromatic light, the contrast of the fringes is very high but surface steps of neighboring pixels of more than one quarter of the wavelength are not displayed correctly. Therefore, this method is not usable for rough, e.g. unpolished, surfaces with strong and sudden height variations. VSI uses a white light source. Here, highest intensity interferences (sharpest fringes) are evaluated, which only appear at the best focus position. By changing the focal point, similar height line can be obtained over a larger heights range. The disadvantage is the resolution in the nanometer range, compared to the sub-nanometer range using PSI. For the experiments performed here, an optical profilometer type Veeco WYKO NT110 in the VSI mode was used.

4.3. STRUCTURAL CHARACTERIZATION

Neutron and x-ray diffraction methods were applied to characterize the structure of some single crystalline samples within ingot Berlin004, 005, and 054. The effects of surface deformation on twinning stresses were studied with x-ray diffraction. Supporting information about reciprocal space, scattering at periodic structures, and the Bragg condition is given in Appendix D, section 10.5.1.

4.3.1. DETERMINATION OF LATTICE PARAMETERS

To determine the lattice parameters of powders and polycrystalline samples, a standard test setup can be used. For powders and polycrystalline samples with randomly orientated crystallites, diffracted beams of any plane *hkl* form a diffraction cone (Figure 4.11). Therefore, intensity can be detected without the need of sample alignment or sample rotations, very easily. This is in opposite to the characterization of single crystalline samples. One set of *hkl* planes does not diffract in a cone but only in one distinct direction. Therefore, the single crystal needs to be aligned so that the diffracted beam can be recorded with a detector. Additionally, the sample has to be rotated around one crystallographic axis to determine a set of lattice parameters. To reduce the time of the experiment, single crystal neutron diffraction experiments are often performed with area detectors. With the help of the measured angle of the diffractions, the *d*-spacing can be determined using Bragg's law, since wavelength and diffraction angles are known.

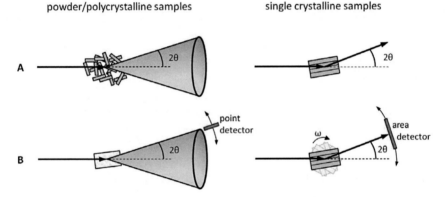

Figure 4.11: Schematic of (A) sample type and (B) experimental setup for powder/polycrystalline sample (left column) and single crystalline samples (right column). The incoming beam gets diffracted in forms of a cone in the powder and polycrystalline sample, so that a moving point detector is sufficient to record all reflections. Single crystalline samples diffract in a specific direction so that the sample needs to be aligned and for each 2θ position of the area detector, the sample needs to be rotated around its axis.

4.3.2. SINGLE CRYSTAL NEUTRON DIFFRACTION

4.3.2.1. BENSC E3

The lattice parameters of single crystalline Ni-Mn-Ga samples were determined at the E3 beamline [83] at the Berlin Neutron Scattering Centre (BENSC) of the HZB, by measuring the Bragg reflections of one hk0 plane and using Bragg's law. E3 is a triple axes diffractometer and is designed for residual stress measurements, providing a 2θ resolution of 0.04°. Furthermore, E3 can also be used as single crystal diffractometer.

The instrument E3 provides a 2θ angle range between 35° and 110° and an ω range of 270° and is equipped with a double focussing bended silicon monochromator. The sample environment includes a Eulerian cradle or cryostat and furnace. A 2-dimensional area detector (300 mm x 300 mm) with 256 x 256 pixels covers a diffraction range of approximately 10°. The samples are probed with a monochromatic beam with a wavelength of 1.486 Å and flux of 5 x 10^6 cm^{-2}s^{-1}. The instrument is equipped with an optical alignment system, which can be used to align the sample on the transition table that can be moved in all three directions and tilted around two axes. This allows measurements of $hk0$ planes of single crystalline samples by rotating the sample around ω (defined in Figure 4.11) for every particular 2θ range that the detector covers. A schematic of instrument E3 is given in Figure 4.12a and a photograph of the E3 setup for an $hk0$ plane scan is given in Figure 4.12b.

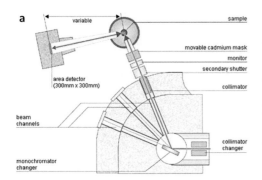 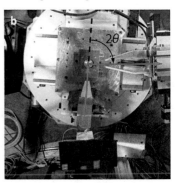

Figure 4.12: (left) Schematic of instrument E3 of the HZB (a). Photograph of the E3 setup from top for single crystal hk0-plane scan (b).

4.3.2.2.   BENSC E2

The instrument E2 is designed for powder and single crystal studies. Because of the large position sensitive detector, low background even at small scattering angles and good resolution, studies of magnetic structures and crystal structures, phase transitions and also diffuse scattering arising from magnetic or structural disorder can be performed at the E2. The detector can be tilted to determine scattering distributions in three dimensions of reciprocal space (flat-cone) with the possibility of energy analysis. These capabilities enable various single crystal applications. Here, E2 was used to investigate the crystal structure of several single crystalline specimens throughout several ingots.

Four 2-dimensional detectors cover a 2θ range of 40° simultaneously but with 10° gaps between each detector. Moving the detector bank by 10°, a total of 80° is covered in two steps. The detectors have similar specification as the E3 detectors of 300 mm x 300 mm and 256 x 256 pixels. The sample stage can be moved, rotated and equipped with different sample environments such as instrument E3. A schematic of instrument E2 is given in Figure 4.13.

During the tests at E3, the analyzer crystals (Figure 4.13) were not installed. The wavelength was for most diffraction experiments 2.390 Å (paralytic graphite monochromator) and for one more detailed test 1.212 Å using a germanium monochromator.

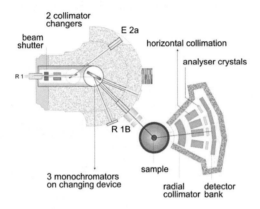

Figure 4.13: Schematic of the instrument E2 of the Helmholtz Centre Berlin[3].

4.4. MAGNETO-MECHANICAL CHARACTERIZATION

In magneto-mechanic experiments MSMA were exposed to a rotating magnetic field and the magnetic-field-induced strain was measured. Figure 4.14 shows a schematic of these experiments and the orientation of the MSMA samples in the three testing devices described in paragraph 4.4.1, 4.4.2, and 4.4.3. The coordination system of the testing devices depends on the orientation of the sample, shown in Figure 4.14. Figure 4.14 shows the two orientations in which the MSMA samples were tested in the magneto-mechanical test devices. The magnetic field is rotating in the y-z plane, i.e. around the x axis of the sample in (a), while the magnetic field is rotating in the x-z plane, i.e. around the y axis of the sample in (b).

[3] http://www.helmholtz-berlin.de/media/media/grossgeraete/
nutzerdienst/neutronen/instrumente/inst/bensc_e2.pdf

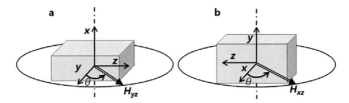

Figure 4.14: Orientations of the MSMA samples in the magneto-mechanical testing devices. Two orientations are shown with the sample orientated so that the magnetic field is rotating around the x axis (H_{yz}) shown in (a) and the sample orientated so that the magnetic field is rotating around the y axis (H_{xz}) in (b).

4.4.1. STATIC MAGNETO-MECHANICAL TESTING DEVICE

The Static Magneto-Mechanical Testing Device (SMMT) is a modified Zwick 1445 (Zwick, Ulm, Germany [84]) mechanical test bench that was donated by the ETH Zurich, Switzerland. The minimum constant cross-head speed is 0.125 mm/min and the maximum load for the used load cell is 500 N ((Burster, type 8432-500). A magnetic field with up to 2 T is available in every direction in the plane normal to the deformation direction. Different setups can be used to examine different magneto-mechanical properties of MSMA. In [22], the SMMT was used to measure the magnetic-field-induced strain in a slow rotating magnetic field. In this work, the SMMT was used to mechanically and magneto-mechanically train and test MSMA single crystal specimens. The SMMT in this setup is shown in Figure 4.15 with lowered (a) and lifted magnet (b). When the magnet is lowered, a magnetic field cannot be applied. Only when the magnet is lifted, a magnetic field of up to 2 T can be applied at any angle in the plane perpendicular to the deformation direction.

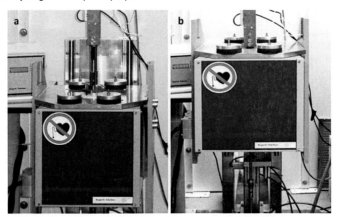

Figure 4.15: Photos of the SMMT in the mechanically and magneto-mechanically setup. In (a) the magnet (blue cube) is in its low position and the magnetic field is not applied, while in (b) the magnet is lifted and a magnetic field can be applied.

Figure 4.16 shows a (a) schematic and (b) photo of the tension/compression inverter of the SMMT. A sample (1) can be deformed compressively between two pistons (2), through which a force is applied parallel to the centerline of the sample (dash-dotted line). The compressive force is produced when the cross head is moving upwards and closing the two interlocked Tungsten grips of the tension/compression inverter. The positions of each piston is transmitted by two quartz rods (3) to two Heidenhain extensometers (type 1281MT). Because the positions of the surfaces of both pistons itself are measured, the error do to elastic deformation of the clamping apparatus is negligible. A magnetic field (blue arrow) can be applied in all directions in the plane perpendicular to the applied force (red arrows). With the magnetic field applied, the MFIS upon unloading and also the magneto-stress can be determined. Because all parts, excluding the sample itself, are non-magnetic, there is no magnetic response of the sample holder on the sample environment.

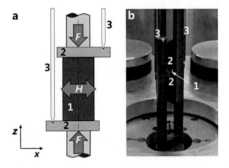

Figure 4.16: Schematic (a) and photo (b) of the SMMT in the mechanically and magneto-mechanically test setup. (1) sample is deformed compressively between two surfaces (2) at which a force is applied parallel to the centerline of the sample (dot-dashed line). The difference between the lower and upper surface position, which is changing during deformation, is transmitted by (3) two quartz rods to two Heidenhain extensometers. A magnetic field (blue arrow) can be applied in all directions in the plane perpendicular to the applied force (red arrows).

The accuracy of the strain measurements is determined by the smallest increment, which the Heidenhain extensiometers can interpolate digitally. The resolution of the SMMT was better than 0.5 N and 20 nm, which resulted in a stress and strain error of less than 0.1 MPa and 3×10^{-6}, respectively, for a standard sample of size 6 mm x 4 mm x 3 mm.

4.4.2. DYNAMIC MAGNETO-MECHANICAL TESTING DEVICE (DMMT)

Some samples were exposed to a rotating magnetic field in the DMMT (Figure 4.17), donated to BSU by the ETH, Zürich, Switzerland. The field of $\mu_0 H = 0.97$ T is produced by a Hallbach cylinder which can be rotated with up to 12,000 rpm. The coordinate system is the sample coordinate system as defined in Figure 4.14. The field was rotated in the x-z or y-z plane, depending on the sample orientation. The elongation of the sample in z direction,

Δz, is transmitted via a push rod to a displacement parallel to the rotation axis outside of the magnet bore. The axial displacement is measured with a resolution of 10 nm with extensometers (Heidenhain, Traunreut, Germany). An error of 10 nm in the deformation measurement corresponds to a relative measuring error of 15×10^{-4} for magnetic field induced deformation of 0.36 mm, which a 6% MFIS of a 6 mm long sample.

Figure 4.17: Photograph of the magneto-dynamic test stand and a close-up of the sensor that is measuring the strain (schematics, see Figure 4.18)

The original setup as provided by ETH Zurich was upgraded at the beginning of this work a new sample holder, heating cooling possibility and optical observation setup (Figure 4.18). The sample (1) is, as before, attached to a sample holder (3) and on the other end to a sliding head (2), which can move when the sample expands and shrinks. The motion of the sliding head is redirected through a ceramic rod (6) and redirecting mechanism (7), so that the motion in z direction can be picked up with a Heidenhain extensiometers. The new setup provides an additional guiding of the sliding head, so that movements to the sides (y direction) and to the top (x direction) is limited. With gaps of approximately 5 µm between the sliding head (2) and the lid (5) as well as between the sliding head and the sides, the displacement error in z direction was reduced significantly in comparison to an open sample holder system without guides. A systematic absolute error due to a possible motion of the sliding head in unwanted directions (x and y direction) is increased by a geometrical factor due to redirecting the z displacement into a vertical movement. This systematic absolute error can therefore still add up to 55 µm, which is an 14% relative error for a sample that deforms by 6%. The new sample holder and sliding head as well as guides and lid were fabricated with Vespel® instead of the machineable ceramic Macor®. Vespel® is self-lubricating and ensures minimal friction during magneto-mechanical testing. The confined sample holder environment as well as the changed material also ensures that a breaking sample does not destroy the sample environment and the fragile ceramic rod.

The other two upgrades that use the new sample holder are schematically shown in Figure 4.18: (a) a heating-cooling setup and (b) general optical observation capability. The heating-cooling upgrade, was first employed for magneto-mechanical testing of Ni-Mn-Ga MSMA foam samples [24]. For these tests, heated and cooled air was directed with a copper tube (8) into the test chamber towards the lid. Conductive heating resulted in the temperature change of the sample, which was measured with a thermocouple (9), which was attached to the sample. This setup allows in-situ MFIS measurements at variable temperature between 0 °C and 60 °C. This is of special interest for samples that go through a phase transformation in this temperature regime.

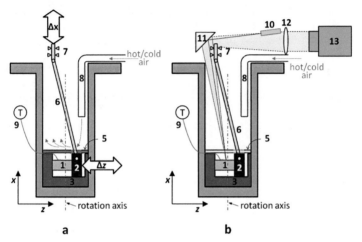

a   b

Figure 4.18: Two experimental setups of the Dynamical Magneto-Mechanical Testing Device (DMMT). In the fatigue setup with temperature control (a), the sample (1) is glued to the sliding head (2) and sample holder (3). The sample holder is attached to the tube (4), which is placed in the center of a Hallbach cylinder, not shown). The sliding head is guided to the sides (y direction) as well as by a lid (5), so that the sliding head can only move in z direction. With the ceramic pushing rod (6) and a redirection mechanism (7), the elongation of the sample in z direction is transmitted to an x movement detectable outside of the Hallbach cylinder with Heidenhain extensio-meters (not shown). All parts except four tiny screws and the sample itself are non-metallic. In the thermo-magneto-mechanical setup (b), a tube (8) directs cooled or heated air into the test chamber. To monitor the samples temperature, a thermo-couple (9) is mounted to the sample surface. The entire setup is placed inside the magnet bore. When optical control of gluing or sample deformation is needed, a focused high powered LED lamp (10) is installed and the sample chamber lid was exchanged with a clear lid. The light of the lamp is redirected via a prism (11) towards the sample surface. A camera (13) with several lenses (12) was used to observe the illuminated surface.

Finally, the DMMT was upgraded with an optical system (schematic shown in Figure 4.18b), with which parts of the sample can be observed during the magneto-mechanical and also thermo-magneto-mechanical tests. The optical system is set up in a way that the hot and cooled air can still be applied. A prism is used to redirect the light of a high powered focused LED lamp towards the sample through a clear lid as well as to create an optical access point for the camera. Due to the confined space, the extensiometers, copper tube and most importantly the ceramic rod, only about 50% of the sample can be imaged. The image quality is also limited because of reflections created on the clear lid and moisture that might develop and condense on this lid during heating and cooling experiments. The optical addition is only intended to confirm proper operation of the experiment for example to ensure that the gluing condition of the sample is maintained during an experiment. The low image quality does not allow to study the initiation and propagation of cracks.

4.4.3. OPTICAL MAGNETO-MECHANICAL TESTING DEVICE (OMMD)

To observe twin boundary motion during magnetically induced deformation at different times of magneto-mechanical training in the DMMT, as well as to observe crack initialization and propagation during cycling and the influence of constraints on the deformation and twin boundary motion, the Optical Magneto-Mechanical Testing Device (OMMD) was built. The electromagnet of the VSM (section 4.2.3) provided the magnetic field. Figure 4.19 shows a schematic of the OMMD.

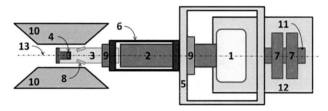

Figure 4.19: Schematic of the optical magneto-mechanical device (OMMD). The camera (1) and optics (2) are mounted with a bracket (9) to an aluminum cage (5). The cage is attached to a hollow shaft (11) which is press fitted in two bearings in the stands (7). The optics are centered along the rotation axis (13) of the shaft. The sample tube (3) that holds the sample holder and sample (4) is mounted to the camera cage (5) by the extension (6) and another clamp (9). The sample surface is illuminated by LEDs (8) placed in the sample tube. The entire OMMD is placed between the pole pieces (10) of the electromagnet on mounting plate (12).

The camera (1, lumera scientific Inc., type Infinity X) and optics (2) are mounted with a bracket (9) to an aluminum cage (5). The cage (5) is attached to a hollow shaft (11), which can hold power and interface cables. The shaft itself is press fitted into two bearings, which are placed in the stands (7). The optics are centered along the rotation axis of the shaft. The sample tube (3), which holds the sample holder and sample (4) is mounted to the camera cage (5) by the extension (6) and another clamp (9). The position of the sample can be adjusted in three directions to focus the sample and to center the sample in front of the camera. The sample surface is illuminated by LEDs (8) placed in the sample tube. The entire OMMD is placed between the pole pieces (10) of the electromagnet (section 4.2.3) and can be rotated around the center axis (13). The different clamps and adjustment of the sample tube position ensures that the sample is stationary relative to the camera.

4.4.4. MANUAL (MAGNETO-)MECHANICAL TRAINING DEVICES (MMT/MMMT)

Due to possible activation of materials exposed to neutron radiation, require magneto-mechanical test devices that can be kept in the experimental hall of BENSC to decontaminated for a longer time if necessary. Additionally, changes in sample deformation have to be quick and exact, to not waste beamtime but also to not lag accuracy. Therefore, two different training devices for the use within the E-Hall were built. The first device was a simple manual mechanical training device (MMT) with which a samples can be deformed in compression without applying torsion while at the same time the deformation can be measured. The MMT (schematic see Figure 4.20) consists of a micrometer (2) and a sleeve (6) on top of the rotating spindle shaft (4) with a flattened stainless steel ball (5). The sample (1) is placed between the flattened ball and the fixed shaft (3). The sample can be deformed in compression by rotating the micrometer spindle that is normally used to tighten the micrometer surfaces to the measured object. The ball is preventing the torsion of the rotating spindle to be transferred to the sample.

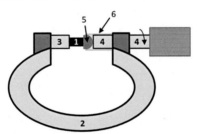

Figure 4.20: Schematic of the manual mechanical training device (MMT). The sample (1) is place in a modified micrometer (2). The sample is put against the fixed shaft (3) of the micrometer. A hardened and flattened metal ball is placed in a sleeve (6) that is centering and holding the metal ball on the rotating shaft (4) of the micrometer. This setup was designed to prevent torsion being transferred to the sample (1) when being compressively deformed.

The second device (Figure 4.21) utilizes a magnetic field perpendicular to the loading direction and is therefore specified as a Manual Magneto-Mechanical Training (MMMT) device. In the MMMT, the sample (3) is clamped between a micrometer (2) with torsion free spindle and a sample pocket (4). The magnets produce a straight magnetic field in the volume of the sample and can be separated in a controlled way so that the magnetic field lines stay parallel. The separated state of the MMMT is shown in Figure 4.21a, photograph with loaded sample is shown in Figure 4.21b and a schematic in Figure 4.21c. In the loaded state, the sample can be deformed in compression using the micrometer and the deformation can be measured. When releasing the micrometer spindle, the applied magnetic field restores the shape and the sample expands perpendicular to the magnetic field.

Figure 4.21: Manual Magneto-Mechanical Training Device (MMMT). (a) and (b) show photographs of the device in a separated (a) and closed (b) condition. (c) is a schematic of the device. The magnets (1) of the MMMT are placed next to the sample (3) which is clamed between the micrometer with torsion free spindle (2) and the sample holder (4). The magnets produce a magnet field perpendicular to the clamping direction. The magnets can be separated in a straight fashion so that the magnetic field lines stay parallel in the center of the sample.

4.5. TRAINING METHODS

4.5.1. THERMO-MECHANICAL TRAINING

Heat treatments are necessary to bias the preferred formation of one twin domain (see chapter 3.5.1). During the thermo-mechanical treatment, force is used as bias. Figure 4.22(a) displays the coordinate system defined on the samples. The shortest edge of the sample was chosen as x axis, the longest edge as the z axis, and the medium edge as the y axis. Figure 4.22 (b) shows the apparatus was used to apply a force during the heat treatment. The sample was placed in the treatment holder (Figure 4.22 b) which was placed in a Fisher Scientific Isotemp® Muffle Oven or on a heating plate. The sample was heated above the martensitic transformation temperature TM while a force was applied to the sample parallel to its x or y direction (Figure 4.22). The force was not released until the sample was completely cooled to room temperature. The temperature during the heat treatment was measured with a thermo-couple, which was attached to the brass block (5) directly next to the sample (1). To induce the force on the sample, a stainless steel bar, Figure 4.22 (3), was bent over the half round brass block (2), which ensured a homogenous loading. By changing the position of the screws (4), the load could be adjusted to a desired stress. To approximate the force on the sample three point bending condition were assumed. The relative error of stress is 15%.

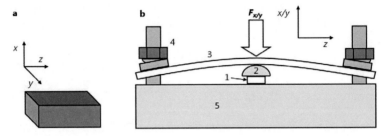

Figure 4.22: Sample geometry and thermo-mechanical treatment. (a) shows the coordinates on the Ni-Mn-Ga sample, x is parallel to the shortest, y parallel to the intermediate, and z parallel to the longest edge; (b) schematic of the apparatus used to apply a force on the sample (1) during the heat treatment. The bending stiffness of the beam (3) was small to minimize relaxation caused by the deformation of the sample. The load was induced and adjusted via screws at the ends of the beam (4). The half round pressure block (2) ensured a homogeneous loading ($F_{x/y}$) of the sample (1). The samples were oriented in x or y direction.

4.5.2. Magneto-Mechanical Cycling and Training

Magneto-mechanical training is based on the main properties of MSMA: twin boundary motion induced by a magnetic field (magneto) which results in deformation. There are two different magneto-mechanical training methods: the first uses the DMMT and its rotating magnet field (cycling). This leads to an activation of twin boundary motion and an exchange of the a and c axes. An ac-twinning system will be established after a number of cycles, which depends on the sample and on constraints [85].

A second magneto-mechanical training method is the compression of a sample perpendicular to an external magnetic field. This training was performed using the SMMT and MMMT. In both devices, the c axis is aligned in a first step by an applied mechanical stress. In a second step, the mechanical stress is released and an external magnetic field perpendicular to the mechanical stress aligns the c axis parallel to the magnetic field. With this training method an ac-twinning system can be created if twinning stresses are low enough.

4.5.3. Mechanical Training: Mechanical Softening

Magneto-mechanical training might not lead to any success, if the twinning stress is so high that deformation cannot be induced with an magnetic field. A method to reduce high twinning stresses, is repeated mechanical deformation in two directions. The c axis is aligned in a first step in one direction of the sample, and in a second step, in a perpendicular direction. These two steps will be repeated as needed. To find the right sample axes that create an ac-twinning system, the sample dimensions during the mechanical softening need to be compared to the maximal possible a/c-ratio. In most cases, the sample axis that expands the most during the initial mechanical deformation, can be compressed in the second step. If then the direction that was compressed during the first the first step is expanding again by the same percentage the the two directions that build an ac-twinning system are found. In the case, where the initially compressed direction is not expanding, a deformation parallel to the third sample axis has to be performed. By comparing the sample dimensions of all three deformation, the set of sample axis that need to be repeatedly compressively deformed, can be found. The repeated deformation will in most cases lead to a reduction of the twinning stress.

4.5.4. THERMO-MAGNETIC TRAINING

Mechanical and magneto-mechanical make use of a mechanical stress. These methods can be applied to bulk samples, but might destroy MSMA samples in a different form, such as foams or wires. A furhter method to align the c axis is thermo-magnetic training. Here, the sample is heated above the martensitic phase transformation temperature and then, while a magnetic field is applied, cooled through the martensite phase transformation. In the ferromagnetic austenite phase, the magnetic moments are aligned easily. If there are no mechanical constraints, this alignment of the magnetic moments, will help aligning the c axis of the martensite. This training method was first employed for the training of foam and polycrystalline bulk samples [23] in a magnetic annealing oven, and later for foam samples in the VSM as described in section 4.2.3. When using the VSM for this training, the reorientation of the c axis can be observed by a changing magnetization of the sample. If training is successful, the magnetization of the sample increases in the direction of the external magnetic field.

4.5.5. THERMO-MAGNETO-MECHANICAL CYCLING

The combination of magneto-mechanical cycling and thermo-magnetic training can be realized by the modified DMMT with the heating-cooling addition. The sample is heated to the austenite phase and then cooled again to the martensite phase while an external magnetic field is constantly rotating around the sample. This method is especially useful for polycrystalline samples and samples with high twinning stresses. This method was first employed during the training of foam samples [24]. The DMMT (section 4.4.2) with the heating/cooling setup (see Figure 4.18b) was used for thermo-magneto-mechanical cycling.

5. OVERVIEW OF SAMPLES AND PERFORMED EXPERIMENTS

5.1. OVERVIEW OF SAMPLES

In this dissertation, samples grown with the SLARE technique were investigated. Three single crystal ingots were grown with nominal compositions of $Ni_{49.0}Mn_{30.0}Ga_{21.0}$ for ingots Berlin004 and Berlin005 and with $Ni_{49.7}Mn_{29.3}Ga_{21.0}$ for ingot Berlin054. Figure 5.1 gives an overview of the position of the slices cut (see section 4.1.3) out of the grown single crystal ingots, as well as the amount of samples that were from each slice. Each slice is indicated by a number (e.g. 01) in the sample name, while each sample cut from this slice receive a designate letter (e.g. 01A is sample A of slice 01). Though sample positions within the slice are unknown, samples properties were tested for one sample per slice, e.g. for all Berlin004 A samples, which is then called sample column A of ingot Berlin004. One column is schematically shown in Figure 5.1 for ingot Berlin004. The naming of the slices was done by MaTecK, which cut the slices.

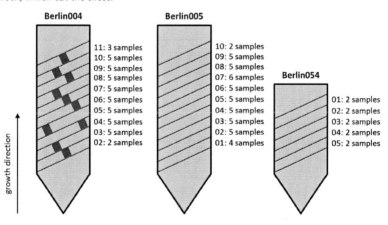

Figure 5.1: Position and amount of samples per slice of single crystal ingots Berlin004, Berlin005, and Berlin054. The dark grey squares indicate exemplarily one sample column.

Since all samples shown were electropolished, tests regarding the influence of surface treatments on twinning stresses and twinning design have been performed on seven additional rest pieces of ingot Berlin005 named A1-A4 and B1-B3. The position of the rest pieces within the ingot as well as the position within one slice is unknown. Since one to two sample columns per ingot were reserved or given to collaborators, so that three samples columns of ingots Berlin004 (named A, C, and E) and Berlin005 (named A, B, and C) and one sample column of ingot Berlin054 (named A) have been used for testing in this dissertation.

To compare magneto-mechanical properties especially over a large amount of magneto-mechanical cycles with samples of the Berlin ingots, one sample from the ETH Zürich, Switzerland, (G31SC100B, short G31B) and one sample from the Universidad del País Vasco, Bilbao, Spain, (Bilbao001B, short B001B) have been examined. Both single crystal samples have a 10M martensite structure, a Curie temperature of about 85 °C (G31B) and 100 °C (B001A), a martensite phase transformation temperature of about 30 °C, and nominal compositions of $Ni_{52.0}Mn_{24.4}Ga_{23.6}$ (G31B) and $Ni_{49.4}Mn_{27.8}Ga_{22.8}$ (B001A).

5.2. ORDER OF EXPERIMENTS

For a systematic characterization of MSMA properties (martensite phase, lattice parameters, chemical composition, transformation and Curie temperature, magnetic anisotropy and saturation magnetization, twinning stress and twinning stress range, mechanically induced strain), sample of every slice or every other slice depending on the experiment of column C of ingot Berlin004 and Berlin005 as well as column A of ingot Berlin054 were studied. All samples of column A and E of ingot Berlin004 received a thermo-mechanical training or were in a selfaccommodated state when their cyclic behavior over 1 million magneto-mechanical cycles was tested. All Berlin005 samples of column A and B were repeatedly thermo-mechanically deformed and repeatedly mechanically deformed (mechanical softening). Those samples that had a low twinning stress were tested for their high cycle magneto-mechanical behavior.

Finally, as indicated above, the twinning stress of seven unpolished rest pieces (A1-A4 and B1-B3) of single crystal ingot Berlin005, were characterized with regards to surface roughness (or surface deformation) by first reducing the surface roughness and then increasing the surface roughness again (sample A1-A4). Samples B1-B3 were used as reference samples. The samples B1-B3 were not polished but repeatedly deformed as samples A1-A4 necessary for twin stress characterization.

Furthermore, phase transformation temperatures and Curie temperatures have been measured for all Berlin004, 005, and 054 samples have been measured with the DSC. The chemical composition has been measured with EDX additionally for Berlin004 column B and Berlin005 column D.

6. INFLUENCE OF GROWTH POSITION ON PROPERTIES OF NI-MN-GA MSMA

6.1. INTRODUCTION

In this part of the dissertation, the characteristic properties of samples of ingots Berlin004, 005, and 054 were systematically investigated along the growth direction of the ingots. These properties include composition, valence electron density (calculated), martensitic phase transformation temperature, Curie temperature, lattice parameters, martensite structure, magnetization curves for all <100> type lattice directions, magnetic anisotropy, and saturation magnetization. Additionally, the twinning stress and development of twinning stress over four consecutive 2-dimensional compressive deformation cycles (each cycle consists of deformations in two directions) were investigated. With x-ray tomography, the pore density and pre distribution throughout the ingot was examined.

To the author's knowledge, this is the most comprehensive overview of Ni-Mn-Ga single crystal sample properties to date. Furthermore, this overview was performed for samples of three single crystal ingots, two with similar nominal composition (Berlin004 and Berlin005) and one ingot with deviating composition (Berlin054) as described in chapter 5.1. While the nominal composition of ingots Berlin004 and 005 were optimized for high martensite transformation temperatures, nominal composition of Berlin054 was chosen to yield higher 10M martensite volume fraction. Similar results without the thermal analysis have been presented by Sozinov et al. [68] for three different single crystals, with the 10M, 14M, and NM martensite structure, are. Schlagel et al. [63] have presented an extensive study of the chemical segregation of three single crystal ingots with different lengths is presented. In [63] the results of the chemical compositions depending on sample position, the solidus, liquidus, martensite start, and Curie temperatures depending on composition have been described. Lanska et al. [87] and Richard et al. [88] have summarized the composition dependence of martensite transformation and Curie temperature as well as martensite structure and lattice parameters for each 33 polycrystalline [87] or single crystalline [88] Ni-Mn-Ga samples. In several publications, the Curie and phase transformation temperature, have been listed in dependence of Ni-Mn-Ga composition [89,90], position on growth axis [91] and alloying elements [92]. Additionally, Babita et al. [93] have compared Curie temperatures and transformation temperatures measured with DSC (heat flow) and VSM (magnetization) as it similarly was done in this dissertation but on crystals grown with different nominal composition. In combination with [63,87,88,89], this chapter builds a comprehensive foundation for further single crystal growth projects aiming to grow 10M, 14M, or NM Ni-Mn-Ga single crystal samples.

6.2. PROCEDURE OF EXPERIMENTS

The samples characterized in this set of experiments were samples Berlin004 03C to 11C, Berlin005 01C to 09C, and Berlin054 01A to 05A. As described before, the naming of sample of the Berlin054 ingot is reverse, since they got named not according to the growth direction but by the order of cutting. Therefore, sample 05A was cut from a slice closest to the single crystal growth starting point and 01A near the end of the growth. The thermal properties and chemical composition of all samples were characterized with DSC (section 4.2.4) and EDX (section 4.2.2). DSC measurements were performed once at 10 °C/min followed by one heating cooling cycle with 5 °C/min, as described in section 4.2.4. The composition of the samples was measured on ten spots (five on each opposing surface) per sample and averaged as described in section 4.2.2.

From now on, only every other sample was characterized in more detail because of limited beam time or instrument access time. These samples were Berlin004 03C, 05C, 07C, 09C, and 11C, Berlin005 01C, 03C, 05C, 07C, and 09C, as well as Belrin054 05A, 03A, and 01A. The lattice parameters and the martensite structure at room temperature was determined with neutron diffraction. Samples Berlin004 03C, 05C, 07C, 09C, and 11C as well as Berlin005 01C, 03C, 05C, 07C, and 09C were examined at beamline E2 of the HZB. Beamline E3 of the HZB was utilized to characterize samples Berlin054 05A, 03A, and 01A. To increase the intensities of diffracted beams to decrease measurement time, the samples were brought into a single domain state by compressing them manually several times at room temperature. For NM samples, which are very hard at room temperature, were thermo-mechanically training (see section 4.5.1). The repeating deformation was stopped when a single crystalline state of the sample was reached as determined from the changes of the sample's dimensions. The rotation axis of the first neutron diffraction scan of each sample, was the axis of compressive deformation. A second neutron diffraction scan was performed around a second sample axis perpendicular to the first one.

Following the neutron diffraction experiments, scanning the samples with x-ray tomography revealed the pore and pore distribution throughout the single crystal ingots. The magnetization was then measured with varying temperature with a heating and cooling rate of 5 °C/min as described in section 4.2.3. The phase transformation produced a selfaccommodated martensite. Based on the size data collected during the training procedure for the neutron experiments, the samples were compressively deformed with the static-magneto-mechanical test device (SMMT, see section 4.4.1). The SMMT tests were performed to determine twinning stresses, change of twinning stresses over several consecutive deformation cycles and maximum mechanical deformation. Samples of ingots Berlin004 and Berlin005 were compressively deformed during four cycles, while samples of ingot Berlin054 were deformed between two to four cycles. The reduction of deformation cycles was necessary to not damage the samples further because cracks have been observed in the samples prior to the tests. Based on the stress-strain behavior (twinning stress and maximum mechanical deformation) the martensite phase present at the moment of the test was identified. The twinning stress, σ_{tw}, is defined here as the mean value of the stress within the defined strain range. The twinning stress was evaluated by graphically integrating the stress values. The difference between the lowest and highest applied stress within these two strain marks define the twinning stress range, $\Delta\sigma_{tw}$. Each deformation cycle consisted of compressive deformation tests along two orthogonal directions. The second deformation direction was chosen to be the one parallel to the sample direction, along which the sample expanded the most during the first compressive deformation test.

For the following magnetic anisotropy measurement, the trained state of the sample resulted from the twinning stress test, which established a nearly single variant state. This was necessary, because the magnetic anisotropy constant can only be determined for a sample in a single variant state. Since twin boundary motion would realign the easy magnetization direction of a soft sample, the samples had to be constraint. The sample holder custom made for these experiments is shown in Figure 4.7.

6.3. RESULTS

6.3.1. BERLIN004

The composition results of samples 03C to 11C of single crystal ingot Berlin004 characterized with EDX are depicted in Figure 6.1a. Along with the composition, the calculated valence electrons density, e/a, corresponding to the determined composition of each sample are presented in Figure 6.1b. While the Ni and Ga content decreased in growth direction from sample 03C with 50.4±0.4 at.-% Ni and 21.7±0.6 at.-% Ga to 11C with 49.0±0.4 at.-% Ni and 20.1±0.3 at.-% Ga, the Mn content increased from 27.9±0.3 at.-% to 30.9±0.3 at.-%. The error bars represent the standard deviations calculated from the ten measured compositions per samples. All electron density values were well within the overlap of all experimental errors. Thus, a tendency was not determined.

The results of the DSC and magnetization versus temperature (VSM) measurements are summarized in Figure 6.2a and b, respectively. Both measurements revealed that the Curie temperature, T_C, decreased in growth direction from 103 °C (03C) to 94 °C (09C) detected with the DSC and from 100 °C (03C) to 92 °C (09C) detected with the VSM. The Curie temperature falls in the range of the martensite transformation for 10C and 11C and could only be measured with VSM, which revealed a higher T_C of sample 11C. T_m increased nearly constantly for both measurement techniques, except for sample 07C measured with the VSM. T_m, increased from 77 °C to 104 °C (DSC) or 69 °C to 95 °C (VSM) in growth direction.

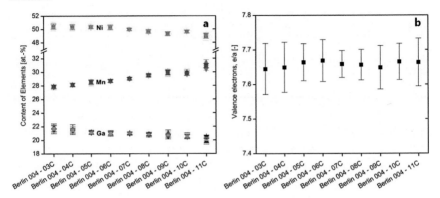

Figure 6.1: Composition determined with EDX (a) and valence electrons (b) for samples Berlin004 C. The Ni content is shown with red squares, the Mn content with blue squares, and the Ga content with green squares. Grey triangles, which are partially overlapping with the colored squares, indicate each element's content for two different surfaces. The error bar of the valence electron density, e/a, was based on the error bars of the composition measurements.

The lattice parameters detected with neutron diffraction and determined martensite structure (from detected martensite modulation reflections) are presented in Figure 6.3 (beamline E2 at the HZB, see section 4.3.2.2). While sample 03C exhibited a 10M martensite phase and 10M modulation reflections, sample 05C and 07C exhibited a 14M martensite phase, and sample 09C and 11C a nonmodulated (NM) martensite phase. The magnitude |1 - c/a| of the spontaneous strain increased with every sample from 0.071 (sample 03C) to 0.222 (sample 11C). While the a and b lattice parameters of sample 03C were equal, they split in samples 05C and 07C before they equalized again in the NM martensite phase of samples 09C and 11C. The values of the lattice parameters are summarized in Appendix C, section 10.4.1.

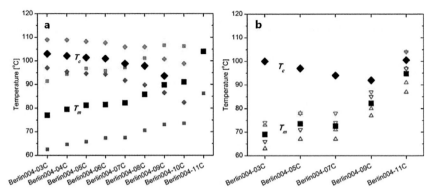

Figure 6.2: Martensite transformation, T_m, and Curie temperature, T_C, characterization of column C of ingot Berlin004. Temperatures were analyzed with DSC (a) and VSM (b). During the DSC measurements, the average Curie temperature (diamond) and the Curie temperature determined upon heating (red diamond), $T_{C,h}$ and upon cooling (blue diamond), $T_{C,c}$ were determined. Similarly, the martensitic phase transformation temperature is the average of the peak temperatures, A_p (heating, read squares) and M_p (cooling, blue squares). During the VSM measurement (b), the Curie temperature (diamond) was determined upon heating (red) and cooling (blue). The austenitic start, A_s (blue triangle pointing up), and finish, A_f (blue triangle pointing down), as well as the martensitic start, M_s (red triangle pointing down), and finish, M_f (red triangle pointing up), temperatures have been determined and T_m from the average of A_f, A_s, M_f, M_s.

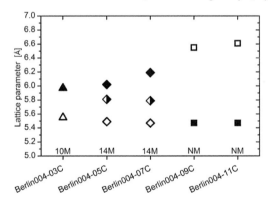

Figure 6.3: Lattice parameters of the martensite phase for samples of ingot Berlin004 at room temperature found with neutron diffraction. The tringles symbolize 10M martensite lattice parameters (open: c, solid: a), the diamonds 14M martensites (open: c, half filled: b, solid: a), and the squares nonmodulated, NM, martensites (solid: a, open: c). The martensite structure based on the observed modulation reflection pattern are given at the bottom of the graph.

The x-ray tomography revealed that there were over 290 pores of size 25 μm and larger in sample 03C, over 180 in 05C, over 190 in 07C, over 230 in 09C, and over 290 in sample 11C. These numbers resulted in different pore densities between 2.5 pores/mm^3 (sample 05C) and 4.0 pores/mm^3 (sample 03C). The change in pore density is not monotonous, but started at a maximum for sample 03C with a minimum for sample 05C and then increased monotonously with crystal position to sample 11C. There was only one crack in samples 03C, 05C, and 07C but six in sample 09C and 17 in sample 11C.

Table 6.1: Results of the x-ray tomography scans of samples 03C, 05C, 07C, 09C, and 11C of ingot Berlin004.

Berlin004		03C	05C	07C	09C	11C
$\rho_{pores > 25 \mu m}$	mm^{-3}	4.0	2.5	2.6	3.1	3.9
$\rho_{pores > 50 \mu m}$	mm^{-3}	0.3	0.2	0.5	0.5	0.4
cracks	-	1	1	1	6	17

Mechanical training over four consecutive compressive deformation cycles that after the initial compression (named loading 1), the twinning stress, σ_{tw}, decreased significantly in all samples showed (see Figure 6.4). For sample 03C, training reduced the twinning stress to 2.4 MPa, for sample 05C to 2.1 MPa, for sample 07C to 7.9 MPa, for sample 09C to 10.7 MPa and for sample 11C to 13.5 MPa. Parallel to the twinning stress, the twinning stress range, $\Delta\sigma_{tw}$, increased from 1.4 MPa for sample 03C to 21.1 MPa for sample 11C. The maximum mechanically induced strain varied from sample to sample and increased from 6.1% (03C), over 10.1% (05C), to 14.4% (07C) before it decreases to 13.8% (09C). Because of the nominal load of the load cell (500N), the deformation experiment of sample 11C had to be interrupted before the slope of the stress-strain increased steeply. The low total strain value of 4.7% of sample 11C which is less than |1-c/a| indicated that twinning was not complete.

The saturation magnetization decreased constantly from 4.96x10^6 A/m (sample 03C) to 3.59x10^6 A/m (sample 11C). The results of the direction dependent magnetization measurements with constraint samples are depicted in Figure 6.4. While sample 03C had one easy and two hard magnetization direction, samples 05C and 07C had additionally an intermediate magnetization direction, and samples 09C and 11C an easy magnetization plane and a hard magnetization direction. The magnetic anisotropy between the easy and hard direction of samples 03C, 05C, and 07C remained fairly constant around 1.7x10^5 J/m^3. The magnetic anisotropy between the easy magnetization plane and hard magnetization direction of sample 09C and 11C was around 2.1x10^5 J/m^3. The magnetic anisotropy between the intermediate and hard magnetization direction of sample 05C and 07C was around 0.75x10^5 J/m^3. The magnetic anisotropy of the directions the easy magnetization plane of sample 09C and 11C is nearly negligible as seen in Figure 6.4. All results of samples of ingot Berlin004 that were investigated here, are summarized in Appendix C, section 10.4.1.

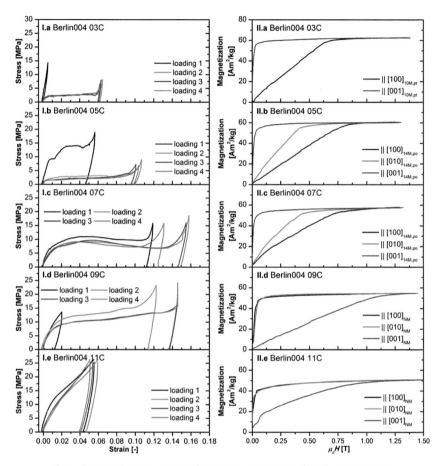

Figure 6.4: Twinning stresses (I) and magnetic anisotropies of single crystal ingot Berlin004 samples. (I) For each sample, stress-strain curves parallel to the longest sample dimension of four 2-dimensional compressive deformations cycles are depicted. (II) The different colors of the magnetic anisotropy measurements indicate different lattice direction along which the magnetization was measured.

6.3.2. BERLIN005

The composition results of samples 01C to 09C of single crystal ingot Berlin005 characterized with EDX are depicted in Figure 6.5a. Along with the composition, the calculated valence electron density, e/a, corresponding to the determined composition of each sample are presented in Figure 6.5b. While the Ni and Ga content decreased in growth direction from sample 01C with 50.7±0.4 at.-% Ni and 21.6±0.3 at.-% Ga to 09C with 49.8±0.3 at.-% Ni and 20.6±0.4 at.-% Ga, the Mn content increased from 27.8±0.2 at.-% to 29.6±0.3 at.-%. The illustrated and named deviations are the standard deviations calculated from the ten measured compositions per samples. All electron density values were well within the overlap of all experimental errors. Thus, a tendency was not determined.

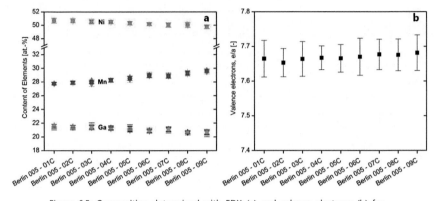

Figure 6.5: Composition determined with EDX (a) and valence electrons (b) for samples Berlin005 C. The Ni content is shown with red squares, the Mn content with blue squares, and the Ga content with green squares. Grey triangles, which are partially overlapping with the colored squares, indicate each element's content for two different surfaces. The error bar of the valence electron density, e/a, was based on the error bars of the composition measurements.

The results of the DSC and magnetization versus temperature (VSM) measurements are summarized in Figure 6.6a and b, respectively. Both measurements revealed that the Curie temperature, T_C stayed constant at 101 °C from sample 01C to 05C before it decreased in growth direction 100 °C (07C) detected with the DSC. A similar behavior was observed with the VSM: T_C stayed constant at 98/97 °C between sample 01C and 05C and then decreased to 94 °C at sample 07C before it increased again to 96 °C in sample 09C. The Curie temperature falls in the range of the martensite transformation for 10C and 11C and could only be measured with VSM, which revealed a higher T_C of samples 07C and 09C. T_m increased constantly for both measurement techniques from 83 °C to 97 °C (DSC) or 78 °C to 88 °C (VSM) in growth direction.

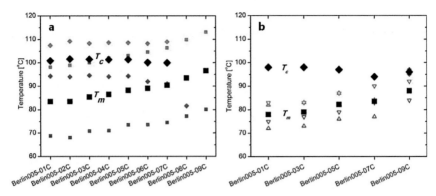

Figure 6.6: Martensite transformation, T_m, and Curie temperature, T_C, characterization of samples Berlin005 C. Temperatures were analyzed with DSC (a) and VSM (b). During the DSC measurements, the average Curie temperature (diamond) and the Curie temperature determined upon heating (red diamond), $T_{C,h}$, and upon cooling (blue diamond), $T_{C,c}$ were determined. Similarly, the martensitic phase transformation temperature is the average of the peak temperatures, A_p (heating, read squares) and M_p (cooling, blue squares). During the VSM measurement (b), the Curie temperature (diamond) was determined upon heating (red) and cooling (blue). The austenitic start, A_s (blue triangle pointing up), and finish, A_f (blue triangle pointing down), as well as the martensitic start, M_s (red triangle pointing down), and finish, M_f (red triangle pointing up), temperatures have been determined and T_m from the average of A_f, A_s, M_f, M_s.

The lattice parameters detected with neutron diffraction and determined martensite structure (from detected martensite modulation reflections) are presented in Figure 6.7 (beamline E2 at the HZB, see section 4.3.2.2). While samples 01C and 03C exhibited a 10M martensite phase and 10M modulation reflections, sample 05C and 07C exhibited a 14M martensite phase, and sample 09C a NM as well as 14M martensite phases. The magnitude |1-c/a| of the spontaneous strain increased with every sample from 0.068 (sample 01C) to 0.22 (sample 09C). The values of the lattice parameters are summarized in Appendix C, section 0.

The x-ray tomography performed on the samples revealed that there were over 230 pores of size 25 μm and larger in sample 01C, over 180 in 03C, over 160 in 05C, over 170 in 07C, and over 220 in sample 09C. These numbers resulted in pore densities between 2.2 pores/mm^3 (sample 05C) and 3.1 pores/mm^3 (sample 01C). As in single crystal ingot Berlin004, the change in pore density in ingot Berlin005 is not linear, but started at a maximum from sample 01C to a minimum at sample 05C and increased again to sample 09C. Besides sample 01C with five visible cracks, all other samples had none or one crack.

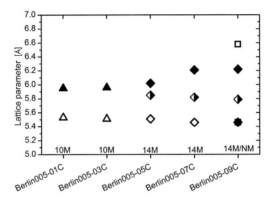

Figure 6.7: Lattice parameters and martensite phase for samples of ingot Berlin005 at room temperature found with neutron diffraction. The tringles symbolize 10M martensite lattice parameters (open: *c*, solid: *a*), the diamonds 14M martensites (open: *c*, half filled: *b*, solid: *a*), and the squares nonmodulated, NM, martensites (solid: *a*, open: *c*). The martensite structure based on the observed modulation reflection pattern are given at the bottom of the graph.

Table 6.2: Results of the x-ray tomography scans of samples 01C, 03C, 05C, 07C, and 09C of ingot Berlin005.

Berlin005		01C	03C	05C	07C	09C
$\rho_{pores > 25 \, \mu m}$	mm^{-3}	3.1	2.5	2.2	2.3	3.0
$\rho_{pores > 50 \, \mu m}$	mm^{-3}	0.3	0.2	0.1	0.2	0.2
cracks	-	5	0	1	0	0

Mechanical training over four consecutive compressive deformation cycles showed that after the initial compression (named loading 1), the twinning stress, σ_{tw}, decreased significantly for all samples (see Figure 6.8). For sample 01C, training reduced the twinning stress to 9.7 MPa, for sample 03C to 1.5 MPa, for sample 05C to 1.0 MPa, for sample 07C to 8.2 MPa and for sample 09C to 14.0 MPa. The twinning range, $\Delta\sigma_{tw}$, decreases and increased with the twinning stress from 9.7 MPa for sample 01C down to 1.2 MPa for sample 05C and up to 23.1 MPa for sample 09C. The maximum mechanically induced strain varied from sample to sample and increased from 4.6% (01C), over 10.3% (03C), to 10.4% (05C) and to 16.1% (07C). Since the maximum stress to create a single variant crystal was not reached upon compressive deformation was not reached for sample 09C, only a maximum mechanically induced strain of 9.7% was reached.

The saturation magnetization decreases constantly from 4.74x10^6 A/m (sample 01C) to 4.03x10^6 A/m (sample 09C). The results of the direction dependent magnetization measurements with constraint samples are depicted in Figure 6.8a. While samples 01C and 03C had an easy and a hard magnetization direction, sample 05C had additionally an intermediate magnetization direction, and samples 07C and 09C an easy magnetization plane and a hard magnetization direction. The magnetic anisotropy between the easy and hard direction of samples 01C, 03C, and 05C remained fairly constant around 1.55×10^5 J/m^3. The magnetic anisotropy between the easy magnetization plane and hard magnetization direction of sample 07C and 09C was around 2.35×10^5 J/m^3. The magnetic anisotropy between the intermediate and hard magnetization direction of sample 05C was around 0.7×10^5 J/m^3. The magnetic anisotropy of the directions of the easy magnetization plane of sample 07C and 09C is negligible as seen in Figure 6.8b. All results of samples of ingot Berlin004 that were investigated here, are summarized in Appendix C, section 10.4.2.

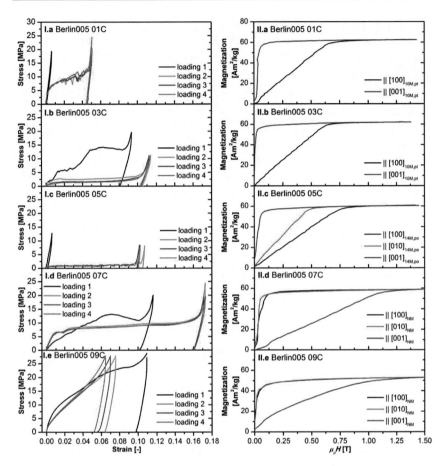

Figure 6.8: Twinning stresses (I) and magnetic anisotropies of single crystal ingot Berlin005 samples. (I) For each sample, stress-strain curves parallel to the longest sample dimension of four 2-dimensional compressive deformations cycles are depicted. (II) The different colors of the magnetic anisotropy measurements indicate different lattice direction along which the magnetization was measured.

6.3.3. BERLIN054

The composition results of samples 05A to 01A of single crystal ingot Berlin054 characterized with EDX are depicted in Figure 6.9a. Along with the composition, the calculated valence electron density, e/a, corresponding to the determined composition of each sample are presented in Figure 6.9b. While the Ni and Ga content decreased in growth direction from sample 5A with 50.7±0.4 at.-% Ni and 23.1±0.3 at.-% Ga to 1A with 49.2±0.5 at.-% Ni and 21.3±0.5 at.-% Ga, the Mn content increased from 26.1±0.3 at.-% to 29.6±0.4 at.-%. The illustrated and named deviations are the standard deviations calculated from the ten measured compositions per samples. All electron density values were well within the overlap of all experimental errors. Thus, a tendency was not determined (Figure 6.9b).

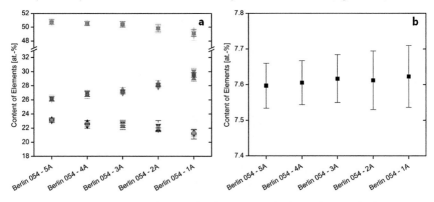

Figure 6.9: Composition determined with EDX (a) and valence electrons (b) for samples Berlin054 A. The Ni content is shown with red squares, the Mn content with blue squares, and the Ga content with green squares. Grey triangles, which are partially overlapping with the colored squares, indicate each element's content for two different surfaces. The error bar of the valence electron density, e/a, was based on the error bars of the composition measurements.

The results of the DSC and magnetization versus temperature (VSM) measurements are summarized in Figure 6.10a and b, respectively. The DSC measurements revealed that the Curie temperature, T_C stayed nearly constant at 100/102 °C from sample 5A to 3A before it decreased in growth direction 93 °C (1A). A similar behavior was observed with the VSM: T_C stayed constant at 103/104 °C between sample 5A and 3A and then decreased to 94 °C at sample 1A. T_m increased constantly for both measurement techniques from 51 °C to 72 °C (DSC) or 43 °C to 58 °C (VSM) in growth direction.

The lattice parameters detected with neutron diffraction and determined martensite structure (from detected martensite modulation reflections) are presented in Figure 6.11 (beamline E2 at the HZB, see section 4.3.2.2). While sample 05A to 03A exhibited a 10M martensite phase and 10M modulation reflections, sample 01A exhibited a NM martensite phase. The deviation of the c/a ratio from 1 was stable for samples 05A to 03A at 0.069 and increased to 0.213 in sample 01A. A 14M martensite phase was not observed in single crystal ingot Berlin 054 with neutron diffraction. The values of the lattice parameters are summarized in Appendix C, section 10.4.3.

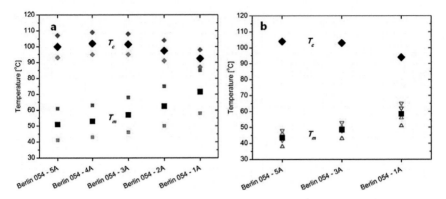

Figure 6.10: Martensite transformation, T_m, and Curie temperature, T_C, characterization of samples Berlin054 A. Temperatures were analyzed with DSC (a) and VSM (b). During the DSC measurements, the average Curie temperature (diamond) and the Curie temperature determined upon heating (red diamond), $T_{C,h}$ and upon cooling (blue diamond), $T_{C,c}$ were determined. Similarly, the martensitic phase transformation temperature is the average of the peak temperatures, A_p (heating, read squares) and M_p (cooling, blue squares). During the VSM measurement (b), the Curie temperature (diamond) was determined upon heating (red) and cooling (blue). The austenitic start, A_s (blue triangle pointing up), and finish, A_f (blue triangle pointing down), as well as the martensitic start, M_s (red triangle pointing down), and finish, M_f (red triangle pointing up), temperatures have been determined and T_m from the average of A_f, A_s, M_f, M_s.

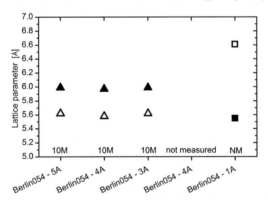

Figure 6.11: Lattice parameters and martensite phase for samples of ingot Berlin005 at room temperature found with neutron diffraction. The tringles symbolize 10M martensite lattice parameters (open: c, solid: a and the squares nonmodulated, NM, martensites (solid: a, open: c). The martensite structure based on the observed modulation reflection pattern are given at the bottom of the graph.

The reconstruction of the x-ray tomography scans for the samples of ingot Berlin054 was performed differently than for ingots Berlin004 and Berlin005. Averaging of background and averaging of the scans itself from which then the background was subtracted resulted in very smooth but unfortunately not as detailed x-ray tomography cross sections like of the other ingots. Therefore, only pores larger than 50 if not 60 µm have been detected. Cracks were invisible even though present in all samples. The number of these larger pores was nearly constant throughout these samples with a pore density of 0.1 to 0.2 pores/mm^3. Results are given in Table 6.3.

Table 6.3: Results of the x-ray tomography scans of samples 05A, 03A, and 01A of ingot Berlin054.

Berlin054		05A	03A	01A
$\rho_{pores > 50\,\mu m}$	mm^{-3}	0.2	0.2	0.1

The mechanical training of the samples over several consecutive compressive deformation cycles showed (see Figure 6.12) that the twinning stress, σ_{tw}, stayed in all samples. The twinning stress for sample 05A was 0.8 MPa, for sample 03A 1.1 MPa, and for sample 01A 6.8 MPa. The twinning range, $\Delta\sigma_{tw}$ was 0.8 MPa for sample 05A, 1.1 for sample 03A and 6.3 MPa for sample 01A. The maximum mechanically was 6.0% (05A), 6.3% (03A), and 15.7% (01A).

The magnetization measurements showed that the saturation magnetization decreases constantly from 5.39×10^6 A/m (sample 05A) to 4.46×10^6 A/m (sample 01A). The results of the direction dependent magnetization measurements with constraint samples are depicted in Figure 6.12a. While samples 05A and 03A had an easy and a hard magnetization direction, sample 01A had an easy magnetization plane and a hard magnetization direction. The magnetic anisotropy between the easy and hard direction of samples 05A and 03A remained fairly constant around 1.85×10^5 J/m^3. The magnetic anisotropy between the easy magnetization plane and hard magnetization direction of sample 01A was at -2.17×10^5 J/m^3. The magnetic anisotropy of the directions of the easy magnetization plane of sample 01A is negligible as seen in Figure 6.12b. All results of samples of ingot Berlin054 that were investigated here, are summarized in Appendix C, section 10.4.3.

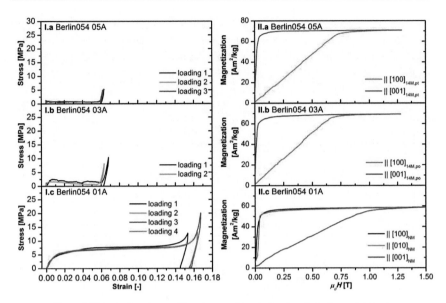

Figure 6.12: Twinning stresses (I) and magnetic anisotropies of single crystal ingot Berlin054 samples. (I) For each sample, stress-strain curves parallel to the longest sample dimension of four 2-dimensional compressive deformations cycles are depicted. (II) The different colors of the magnetic anisotropy measurements indicate different lattice direction along which the magnetization was measured.

6.4. DISCUSSION

The results of section 6.3 give a comprehensive overview of properties of samples of Ni-Mn-Ga single crystal ingots with nominal compositions of $Ni_{49.0}Mn_{30.0}Ga_{21.0}$ for ingots Berlin004 and Berlin005 and $Ni_{49.7}Mn_{29.3}Ga_{21.0}$ for ingot Berlin054.

The martensite phases of all samples are listed in Table 6.4 sorted according to the methods with which they were characterized. Samples with the 10M structure show 10M characteristics in all performed tests, i.e. neutron diffraction (lattice parameters and modulation reflections), compressive deformation (approximately 6% MFIS) and magnetization tests (easy magnetization along c axis and hard magnetization along a and b axes). Samples with the 14M or NM structure, the distinction between those two phases is not that simple. While sample Berlin005 07C showed 14M modulation reflections, the c/a ratio was between the literature values of the 14M and NM structures. Additionally, the stress-strain behavior with elevated twinning stress and mechanically induced strain of 16.1% rather points to a NM structure or phase mixture with both structures. The large strain might also be caused by a stress-induced intermartensitic phase transformation resulting from the high stresses of up to 15 MPa. This assumption is confirmed by the magnetic anisotropy measurements, which were performed after the mechanical tests and revealed an easy magnetization plane and hard magnetization axis characteristic for the NM structures. The example Berlin005 07C shows that all results have to be compared and analyzed taking into account the order of tests and the test specifications.

Figure 6.13a summarizes the martensite structures found in the samples listed in section 6.3. It can be seen that the e/a ratio has very little if any influence on the martensite phase of the samples. A stronger correlation of the martensite structure with the Ni and Mn content exists. At higher Mn (>28.5 at.-%) and lower Ni (<50 at.-%) content, the NM structure is preferred, while at higher Ni (>50.5 at.-%) and lower Mn (<28.0 at.-%) content the 10M phase is present. In the range 14M martensite phase is detected. In Figure 6.13b the Ni/Mn, Ni/Ga, Mn/Ga ratios and the valence electron density, e/a, are given. While the number of valence electrons stays fairly constant, sample properties (e.g. martensite structure) are obviously changing over the length of the single crystal ingot. In contrast to the number of valence electrons, the Ni/Mn and Mn/Ga ratios vary along the crystal ingot. Therefore, all properties of the Ni-Mn-Ga sample will be related to the Mn/Ga ratio, which shows the largest change throughout a single crystal ingot.

Table 6.4: Structure of the samples 03C to 11C of ingot Berlin004, 01C to 09C of ingot Berlin005, and 05A to 01A of ingot Berlin054. 14M-NM indicates the possibility of mixed 14M-NM structures.

	Sample Berlin...	Modulation reflections	Lattice parameter c/a ratio	σ-ε-curves	Magnetic anisotropy
	03C	10M	10M	10M	10M
	05C	14M	14M	14M	14M
004	07C	14M	14M	14M-NM	14M-NM
	09C	NM	NM	14M-NM	NM
	11C	NM	NM	NM	NM
	01C	10M	10M	10M	10M
	03C	10M	10M	14M	10M
005	05C	14M	14M	14M-NM	14M
	07C	14M	14M-NM	14M-NM	NM
	09C	14M-NM	14M-NM	NM	NM
	05A	10M	10M	10M	10M
054	03A	10M	10M	10M	10M
	01A	NM	NM	14M-NM	NM

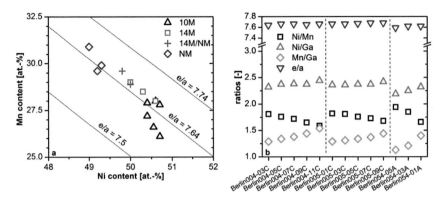

Figure 6.13: Martensite phases depending on Ni and Mn content (a) and ratios (Ni/Mn, Ni/Ga, Mn/Ga, e/a) of all characterized samples (b).

Figure 6.14 summarizes the Curie temperature and martensite transformation temperature as well as the difference between both temperatures as function of the Mn/Ga ratio. In Figure 6.14a, values measured with the VSM are shown in Figure 6.14b measured with the DSC. Small deviation of the actual temperature values are due to a systematic error due to different delays in temperature change rates of both measuring techniques. Besides these deviations, similar trends can be seen. The Curie temperature varies only slightly with the Mn/Ga ratio. At Mn/Ga ratios of 1.1 the Curie temperature is slightly above 100 °C, at ratios of 1.5 slightly below 100 °C. In contrast, the martensitic transformation temperature increases from around 50 °C to approximately 100 °C over the same Mn/Ga ratio range. Thus, the difference between both temperatures decreases from 50 °C to zero.

In Figure 6.15, the lattice parameters and the magnitude of the magnitude $|1-c/a|$ of the spontaneous strain are plotted as function of the Mn/Ga ratio. At Mn/Ga < 1.33, the 10M structure was present in Ni-Mn-Ga single crystals with $|1-c/a|$ < 0.075. For 1.33 < $|1-c/a|$ < 1.45, the 14M structure was found with $|1-c/a|$ values of 0.087 and 0.122. That the NM structure appeared already at Mn/Ga = 1.4 and above, at which the 14M structure was also present. The NM structure exhibited $|1-c/a|$ values of 0.189 to 0.209. Thus, a clear trend from 10M at lower Mn/Ga ratios over 14M and mixed 14M/NM to NM structure was found. The lattice parameters are in good agreement with the literature values [68] with a maximum deviation of ±3%.

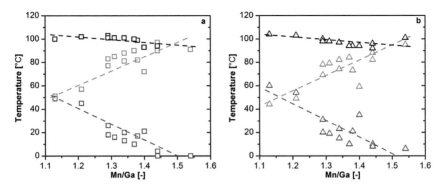

Figure 6.14: Curie temperature (black), martensite transformation temperatures (red), and the difference of both temperatures (blue) over the Mn/Ga ratio. (a) shows the values measured with VSM, (b) these measured with the DSC.

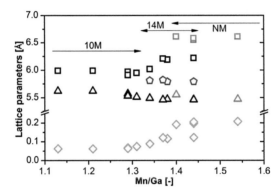

Figure 6.15: Lattice parameters and |c/a-1| over Mn/Ga. For 10M and 14M martensites the black squares symbolize a, the black triangles c, and the blue pentagon b, while for NM martensites the red squares symbolize c and the red triangles a. |c/a-1| is displayed as orange diamonds.

In Figure 6.16, the twinning stress, twinning stress range, and the mechanically induced strain are shown. Since |1-c/a| is the theoretical limit of induced strain produced by twin boundary motion, the mechanically induced strain should increase with increasing Mn/Ga. This was true for Mn/Ga > 1.42. Above this value, the mechanically induced strain was decreasing again. This can be explained with the help of the stress-strain curves recorded for the NM samples. In contrast to the stress-strain curves of 10M and 14M martensites, which exhibit a stress plateau, such a plateau is not present for samples with NM structure (Figure 6.4 Ie, Figure 6.8 Ie, and Figure 6.12c). Rather, the stress increases monotonously with strain. The force limit of 500 N was therefore reached before the maximum strain was produced. Samples with Mn/Ga < 1.35 had with one exception twinning stresses and twinning stress ranges below 2.5 MPa. Increasing the Mn/Ga ratio resulted in increased strain, increased twinning stress and an increased twinning stress range.

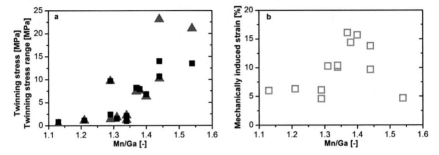

Figure 6.16: (a) Twinning stress σ_{tw} (black squares, left scale), and twinning stress range $\Delta\sigma_{tw}$ (blue triangles), (b) mechanically induced strain (red squares, right scale), versus the Mn/Ga ratio.

The x-ray tomography experiments also revealed an interesting trend for single crystal ingots Berlin004 and Berlin005. At the bottom and the top of the single crystal ingot, pores seem to be present in higher densities than in the center. Additionally it was shown, that ingot Berlin004 had a higher pore density than ingot Berlin005. This might be one of the reasons why twinning stresses are higher in ingot Berlin004 than Berlin005.

Figure 6.17 shows the magnetization and magnetic anisotropy of all samples as a function of Mn/Ga. While only three magnetization measurements along the intermediate magnetization axis of 14M martensites have been measured and a trend is therefore not apparent, it can be seen that K_b is a factor of 2.5 smaller than K_a. The magnetic anisotropy between the easy and hard magnetic axis or planes show a very distinct behavior. Within the 10M phase, K_a is decreasing from around 2×10^5 J/m^3 to 1.4×10^5 J/m^3 with increasing Mn/Ga. In the 14M phase, K_a is increasing again with increasing Mn/Ga to above 2.3×10^5 J/m^3. In the NM phase K_a follows the trend of the 10M phase and decreases with increasing Mn/Ga to 2×10^5 J/m^3. The saturation magnetization decreases linearly from 5.5×10^5 A/m at Mn/Ga $= 1.12$ to 3.6×10^5 A/m at Mn/Ga $= 1.54$. Thus, the change of the saturation magnetization is independent of the martensite structure, while the dependence of the magnetic anisotropy varies with the martensite structure. All magnetic anisotropy values are slightly larger than values found in literature: $K_{a,10M} = 1.45 \times 10^5$ J/m^3, $K_{a,14M} = 1.6 \times 10^5$ J/m^3, $K_{b,14M} = 0.7 \times 10^5$ J/m^3 [68]. The saturation magnetization is on the other hand lower than the literature value of 6×10^5 A/m for all three martensite structures.

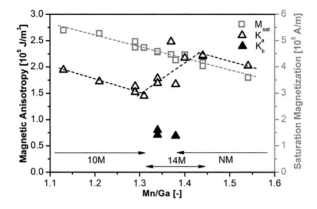

Figure 6.17: Saturation magnetization and magnetic anisotropy over Mn/Ga. The open square indicates the saturation magnetization, the open triangle the magnetic anisotropy between easy and hard magnetization directions, K_a, and the solid triangle the magnetic anisotropy between the intermediate and hard magnetization axes, K_b. At the bottom, arrows indicate the Mn/Ga ratio regions of the here found 10M, 14M, and NM martensite phases.

At the end of this chapter, the properties of the martensite phase are compared with properties reported by Lanska et al. [87] and Richard et al. [88]. In Figure 6.18, the martensite structures of 79 Ni-Mn-Ga samples are related to the Ni and Mn content. A general trend of martensite phase distribution over the shown composition range was found, comparing these results. In relatively high Ni and low Mn regions, the 10M structure is present. With increasing Mn content and decreasing Ga content, a shift to the 14M structure was found. Surprisingly, Lanska et al. found the 14M structure not only at higher Mn content but also in regions where the 10M structure was found in this study. In Lanska et al., an increasing e/a resulted in a shift from 10M over 14M to NM. This trend is much weaker in this study and not the case in the study of Richard et al., where the change from 10M to 14M solely depends on the Ni/Mn ratio. The samples characterized in this dissertation, show a mixture of both dependencies. The shift from 10M to 14M and from 14M to NM takes place with increasing Mn content.

Furthermore, Richard et al. did not report on the NM structure, which have been found in this work at Ni and Mn concentrations, which showed 14M structure in Richard et al. Lanska et al. on the other hand, found NM martensite at Ni concentrations above 51 at.-% and at 47 at.-%. NM martensite was only found at lower Ni concentrations (49.3 at.-% and below). This leads to the assumption that an offset of the composition measurements is distorting the martensite phase regions of each study. Several reasons can be identified for this offset. First, different characterization techniques have been used to determine the chemical composition of the sample (electron probe analyzer in Richard et al., and EDX in

Lanska et al. and this dissertation). Alone the systematic error of different EDX is large enough to shift all results by several at.-%. The EDX measurements for this work were performed on the same EDX with constant settings. Lanska et al. do not provide detailed Information about the EDX parameters. Second, the lattice parameters have been determined on powder (Richard et al.) and single crystal (this work and Lanska et al.). Even though the stress introduced during the production of powder samples were released by annealing in the study of Richard et al., the production and larger surface to volume ratio might have had an influence on the martensite structure of in the powder samples.

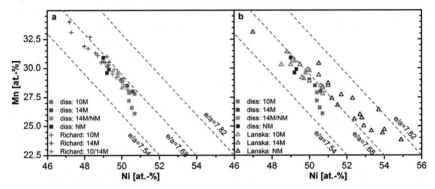

Figure 6.18: Comparison of in this work determined martensite phases (squares) with phases from Richard et al. [88] (crosses) and Lanska et al. [87] (triangles). The colors of the symbols indicate different martensite phases of mixed phases: 10M (red), 10M/14M (pink), 14M (blue), 14M/NM (green), and NM (black).

In this chapter of the dissertation, it was shown that all properties of Ni-Mn-Ga MSMA are highly dependent on the chemical composition. It was further clarified, how the composition also influences the mechanical properties. Finally, the martensite structures found in this study, were compared with two other publications and discrepancies between all studies were found. This shows that further studies on the composition dependency of martensite structure also in lower e/a ratios should be conducted to eventually find application based composition for different operation temperature and magnetic-field-induced strains.

7. INFLUENCE OF SURFACE TREATMENT AND MECHANICAL TRAINING ON THE TWINNING STRESS

7.1. INTRODUCTION

Structure, thermal properties, and magnetic properties of Ni-Mn-Ga magnetic shape memory alloys (MSMAs) are very sensitive to variations of the alloy composition and have been studied in various publications [e.g.53,87,89,94-98]. Similarly, the surface preparation procedures impact the magneto-mechanical properties of these materials. However, information on the surface treatment and sample preparation methods is often of qualitative nature and describes only the cutting and polishing itself [e.g. 8,87,97,99-105]. It is claimed that one reason to do a final electropolishing treatment on bulk Ni-Mn-Ga specimens is the reduction of surface stresses that might be present [103,106].

Thermo-mechanical and thermo-magnetic training increases the magnetic-field-induced strain (MFIS). Murray et al. [11] and Sozinov et al. [107] applied a magnetic field and a mechanical load to create a single variant specimen. This method was applied to achieve high MFIS [108]. The influence of several mechanical treatments as training to produce repeatable magneto-mechanical behavior was mentioned in [68]. In [13], it was shown how variants change from a single variant state to a two variants state after thermo-mechanical training, and how the variants rearrange during straining in different directions and when exposed to a rotating magnetic field. If several variants are present in a single crystalline sample, the application of a magnetic field favors the variant with its c axes most closely parallel to the magnetic field direction. This variant grows at the expense of the other variants, in the ideal case, until a single variant state is attained. Rotating the magnetic field to a different direction favors another variant, which has the c direction close to the direction of the magnetic field causing this variant to nucleate and grow. Rotating the magnetic field creates nearly single variant states with highly mobile twin boundaries with reduced twinning stresses. Additionally, in [14,35,85,109], it is described how training impacts the MFIS at the beginning of a rotating magnetic field experiment and how the MFIS evolves over a large number of magnetic field revolutions. In this part of this study, these previous results [13,14,35,68,85,109] to produce samples with low twinning stress. Additionally, not only one but several samples were trained the same way and only one processing parameter was changed (surface roughness) to examine its influence on the twinning stress. While thermo-mechanical training increased the MFIS at the beginning of the magneto-mechanical cycling experiments in a rotating magnetic field to the theoretical maximum, it also increased the susceptibility to failure by fracture [85,109]. Ineffectively trained samples (with selfaccommodated martensite) have a small MFIS at the beginning of the magneto-mechanical cycling experiments. Due to the rearrangement of twin variants and the growth of favorable twin variants during "in-service training" (i.e. while exposed to a rotating magnetic field), the MFIS increased from 0.18% to above 2% for a sample with 10M martensite without fracturing over 100 million actuation cycles [85]. Straka et al. [110] generated a simple two variant state with a sample with 10M structure by repeated compressive deformation. This lead to a decrease of the twinning stress from 2.3 MPa directly after thermo-mechanical training (cooling from austenite to martensite with applied stress) to 1.1 MPa in a trained state and from 6.3 MPa to 2.9 MPa for a 14M martensite sample.

In this study, which is divided in three test series, it is shown that different surface treatments in combination with repeated mechanical deformation experiments significantly reduce the twinning stress (test series A, also published in [111]). Furthermore, polishing reduces the twinning stress, especially in the high strain regions where the rough surface layer causes hardening, and also leads to serrated flow. In test series B, it is shown that repeated compressive deformation along two sample axes reduces the twinning stress significantly, while repeated deformations along three sample axes does not decrease the twinning stress. In test series C, it is also shown that the increase of surface deformation leads to higher twinning stress. Furthermore, surface strain effectively pin twins and with increasing surface deformation, the twin thickness can be reduced.

7.2. PROCEDURES OF EXPERIMENTS

In test series A, the influence of the surface treatments on the twinning stress of Ni-Mn-Ga single crystalline MSMAs was investigated on nine parallelepiped-shaped samples, which were labeled A1 to A6 and B1 to B3, with faces parallel to $\{100\}_{austenite}$. The influence of 2-dimensional and 3-dimensional compressive deformation experiments with repeated compressive deformation along two (e.g. x-z-x-z...) or three sample directions (e.g. x-y-z-x-y-z...), was studied on samples A2 and A3 in test series B. In test series C, the dependence of the twinning stress on the degree of surface deformation was investigated on samples A1 and A3.

The samples were cut from a single crystal with a nominal composition of $Ni_{49}Mn_{31}Ga_{20}$, which was grown with a modified Bridgeman technique, SLARE (see section 4.1.1). Prior to cutting, the single crystal ingot was heat treated: at 1,273 K for 80 h, at 998 K for 2 h, at 973 K for 10 h, and finally at 773 K for 20 h. The composition of each sample was characterized with electron dispersive x-ray spectrometry (EDS) with a Philips XL 30 ESFM scanning electron microscope (Table 7.1). The martensite transformation temperatures and the Curie temperature were determined from differential scanning calorimetry measurements, which were performed with a Netzsch DSC 404 C calorimeter (Table 7.1). The martensite and Curie temperatures listed in Table 7.1 are the averages of the values obtained during heating and cooling. The Curie temperature varied between 98 and 102 °C. The martensite temperature varied between 77 and 87 °C. The structure characterization was performed on two single crystal specimens (A3 and B3) using neutron diffraction at beamline E3 described in section 4.3.2.1. Because of limited beam time, only two samples were characterized with neutron diffraction. The samples were selected based on a maximum strain of 6% and 10% obtained in deformation experiments. These values indicated that these samples have the 10M and 14M martensite structures, respectively. For the neutron diffraction experiments, sample A3 was in a self-accommodated state, while B3 was tested twice, once in a self-accommodated state and once in a compressed state with a predominant martensite variant.

Table 7.1: Composition, phase transformation temperatures, and structural properties of samples A1-A4 and B1-B3. Because of limited beam time, only samples A3 and B3 were characterized with neutrons. The martensite structures of the other samples were determined from the maximum strain measured in deformation experiments. The maximum strain value was calculated from the shape change of the sample before and after a compressive deformation experiment. The asterisk (martensite structure of B2) indicates that a mixture of 14M and 10M is not confirmed by neutron scattering but assumed due to the maximum strain of 9%.

Sample	Composition [at.-%]			Transf.-Temp. [°C]		Maximum Strain [%]	Martensite Structure
	Ni	Mn	Ga	T_M	T_C		
A1	50.4(3)	28.9(2)	20.7(3)	81	99	10.8	14M
A2	50.2(3)	29.1(4)	20.7(4)	79	98	10.6	14M
A3	50.6(3)	28.3(2)	21.0(3)	77	102	6.0	10M
A4	50.6(2)	28.6(2)	20.8(2)	78	101	6.1	10M
B1	50.8(4)	28.3(2)	20.9(2)	83	99	9.6	14M
B2	50.6(3)	28.2(4)	21.2(4)	83	99	9.0	14M(/10M)*
B3	50.5(2)	28.7(3)	20.8(2)	87	100	8.8	14M/10M

All parallelepiped-shaped specimens were spark eroded from the single crystal ingot. To compare the roughness of different cutting techniques, one cut of the samples A1 to A6 was performed with a precision wire saw (see section 4.1.3.2).

During **test series A**, deformation experiments were performed with all specimens using the SMMT (see section 4.4.1), but only results of samples A1-A4 and B1-B3 are shown here, since samples A5 and A6 had an extremely high twinning stress as it is characteristic for nonmodulated martensites. The samples A1 to A6 were electropolished between deformation experiments until all cutting marks were completely removed, while a set of control samples, B1 to B3, were tested unpolished. The average roughness of each surface was determined with an optical profilometer (see section 4.2.6) before and after each electropolishing treatment. The goal of these deformation experiments was to isolate a single twinning system, such that the samples changed size in only two directions, while the length in the third direction remained constant. Cartesian coordinates were defined on the samples such that the shortest, intermediate and longest edges are parallel to the x, y, and z directions. For the first deformation experiment, each sample was loaded parallel to the longest edge (i.e. in z direction) so that they elongate along the x and y directions, respectively. After this first deformation test, the changes in size of the specimens were measured. Thereby, sample A1 to A4 showed a significant difference in the elongations of x and y. Hence the samples were compressed parallel to the edge that elongated the most in the first deformation experiment. The second deformation, along the y direction for samples A1 and A2 and along the x direction for samples A3 and A4, led to an elongation of the z direction. This resulted in sets of loading directions of y and z for samples A1 and A2, and of x and z for samples A3 and A4. In contrast, the samples B1 to B3 elongated in both directions (x and y) by approximately the same amount. In the second deformation experiment, these samples were loaded in the x direction and in the third experiment in the y direction. During the compression test in y direction, the edge parallel to the x direction expanded the most. This resulted in a set of loading directions for samples B1-B3 of x and y. These sets of deformation experiments constitute one "deformation cycle" of the different samples. After each deformation cycle, the samples A1 to A4 were subjected to a polishing treatment and surface roughness measurement with optical profilometry. The complete sequences of deformation tests are given in Table 7.2.

After the first deformation cycle, samples A1 to A4 were electropolished for 60 s using a 60% sulfuric acid with a current of approximately 0.035 A/mm^2 and a voltage of 7 V. The etch rate was determined after each polishing treatment for each sample by comparing the size before and after polishing. After the first electropolishing treatment of 60 s, a subsequent roughness measurement, and deformation cycle, the specimens A1 to A4 were electropolished for 60 s, then tested, and again electropolished , this time for 120 s before being tested a fourth time. Samples A5 and A6 received the same polishing treatment to reduce the statistical error of the surface roughness measurements. Samples A5 and A6 were not deformed, though.

The largest spark eroded surface and the wire cut surface of samples A1 to A6 were imaged twice with the optical profilometer to evaluate surface roughness. The field of view was chosen such as to lie completely within a single twin domain. In this way, the twin relief did not contribute to the surface roughness. The scanned area was 92 µm x 121 µm for these measurements.

Samples B1 to B3 were not electropolished, but were mechanically tested four times in the same way as the polished samples. After these four deformation cycles, the samples were mechanically polished: first with a 9 µm and then with a 6 µm water-based diamond suspension to remove the surface roughness. Two deformation experiments were performed similarly to the A samples, namely first along the z direction and then along the direction which expanded the most during the first deformation experiment. The complete list of deformation experiments is given in Table 7.2 (deformation cycles 1-4 for the unpolished condition, and cycles 5 and 6 for the polished condition). The average surface roughness of samples B1 to B3 was tested similarly to samples A1 to A6, before and after the mechanical polishing treatments, but on a smaller area (45 μm x 59 μm) because of smaller twin sizes.

Before performing **test series B**, samples A2 and A3 were heated and cooled without applied stress to establish a self accommodated state without prior training as well defined starting state. Then, both samples were subjected to two 2-dimensional deformation cycles with compressive deformations along z and y direction or z and x direction, respectively. Secondly, sample A2 was then exposed to three 3-dimensional deformation cycles in z-y-x direction and sample A3 in a z-x-y-x and a z-x-y deformation cycle. Finally, 2-dimensional deformation cycles followed, i.e. four z-x cycles for sample A3. The 2-dimensional deformation cycle for sample A2 was chosen to be z-y. After the 7th deformation cycle's deformation in z direction, the dimension parallel to the x axis increased and not, as anticipated by the first 2-dimensional tests, the dimension parallel to the y axis. After this deformation, an x-y twinning system was established, and two x-y deformation cycles were performed. The complete sequences of deformation tests are given in Table 7.2.

In **test series C**, samples A1 and A3 remained in their well trained 2-dimensional twinning state. Stress-strain tests were performed on both samples with electropolished surfaces prepared during test series A and after each of the following surface treatments (sequence of tests shown in Table 7.2). The electropolished surfaces (surface treatment 1) were mechanically polished using a 9 μm water based diamond suspension (surface treatment 2), then mechanically grinded with 1200 grit SiC particles (surface treatment 3). Finally, the surfaces were glass ball blasted with four bursts of 0.5 s and a sample glass ball pistol distance of 10 cm (surface treatment 4). Since after surface treatment 1, XRD measurements (see next paragraph) have not been performed, the sample was chemically etched again to characterize the surface without deformed surface layer (surface treatment 5).

The surface roughness of each surface treatment was evaluated on sample A3 on five spots with optical profilometry. The scanned area was 92 μm x 121 μm. Additionally, 2θ and ω scans (rocking curve) of samples A1 and A3 were performed with XRD after surface treatments 2 to 5 during test series C. The Bruker D8 Discover x-ray Diffractometer was configured with a parallel beam setup and equipped with a Copper Kα source and a point detector with copper slits. Experiments had a step size of 0.01° with 1 second per step and a rectangular beam with a width of 1.2 mm. With this XRD setup different source and detector geometries (i.e. $\theta_1 \neq \theta_2$) were used in order to optimize diffraction intensities. Rocking curves were achieved with this setup by changing θ_1 and θ_2 at a constant 2θ with $\theta_1 \neq \theta_2$ around a fixed sample position. The full width half maximum (FWHM) of the 2θ scans is an indicator of elastic strain and residual stress in the surface layer of a sample while the ω scan is an indicator of mosaicity (a measure for crystal rotations) near the surface.

Table 7.2: Sequences of deformation tests. During test series A, samples A1 to A4 were electropolished after each deformation cycle, control samples B1 to B3 were not electropolished between deformation cycles. During test series B, the electropolished samples were investigated, while during test series C, the surface roughness was increased after each mechanical test. The underscores indicate deformation directions for which stress-strain curves are shown in Figure 7.4 and Figure 7.5.

Test series A deformation cycle						
Sample	1	2	3	4	5	6
A1	z̲ y	z̲ y	z̲ y	z̲ y		
A2	z̲ y	z̲ y	z̲ y	z̲ y		
A3	z̲ x	z̲ x	z̲ x	z̲ x	after mechanical	
A4	z̲ x	z̲ x	z̲ x	z̲ x	polishing	
B1	z y̲ x	y̲ x	y̲ x	y̲ x	y̲ x	y̲ x
B2	z y̲ x	y̲ x	y̲ x	y̲ x	y̲ x	y̲ x
B3	z y̲ x	y̲ x	y̲ x	y̲ x	y̲ x	y̲ x

Test series B deformation cycle										
Sample	1	2	3	4	5	6	7	8	9	10
A2	z̲ y	z̲ y	z̲ y	z̲ y x	z̲ y x	z̲ y x	z̲ y	z	x y̲	x y̲
A3	z̲ x	z̲ x	z̲ x y x	z̲ x y	z̲ x	z̲ x	z̲ x	z̲ x	z̲ x	-

Test series C deformation cycle			
Sample	1	2	3	4
A1	z̲ y	z̲ y	z̲ y	z̲ y
A3	z̲ x	z̲ x	z̲ x	z̲ x

7.3. RESULTS

Scanning electron micrographs taken prior to the first polishing treatment of **test series A** show that the topographies of the spark eroded surface and the wire cut surfaces were different and had distinct features (Figure 7.1). The spark eroded surfaces were densely covered with dents and bumps with no flat surface regions between them. These surface features were between 20 and 80 μm wide and had height differences of over 12 μm. The largest dents on the wire cut surfaces measured 30 μm. No salient features were found on these surfaces. The dents only covered a small percentage of the entire surface area. In between the dents, there were flat regions with very small ridges not larger than 1 μm.

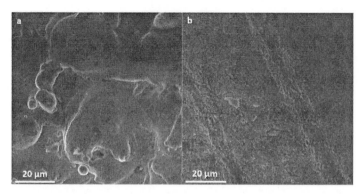

Figure 7.1: Scanning electron micrographs of as cut surfaces. (a) The spark eroded surface showed very large, round surface features in the range of several 10 µm in diameter and depth of up to 12 µm. (b) Features on wire cut surfaces were smaller than 5 µm and had a depth of less than 1 µm.

The images shown in Figure 7.2 were taken with the optical profilometer and illustrate qualitatively how the surface roughness changed with each polishing treatment. After the first polishing treatment, all coarse and rough surface features on surfaces of both cut types were mostly gone and the surface appeared to be smooth. Additionally, straight valleys and ridges at a distance of 2 µm to 20 µm appeared (indicated exemplarily by black dashed lines in Figure 7.2). These large features were twins formed during deformation tests. After the third electropolishing treatment, small cavities with 2 to 5 µm diameter appeared on all surfaces, probably due to pitting.

The effect of material removal on surface roughness is shown quantitatively in Figure 7.3. The average surface roughness, R_a, decreased to less than a third after the first electropolishing treatment. After each following polishing treatment, the surface roughness decreased further, however at a reduced rate. The average material removal during the first, second, and third electropolishing treatment were 29, 14, and 27 µm, i.e. 70 µm were removed totally. This was about 6 times the highest surface feature of the unpolished spark eroded surface. Comparing the average surface roughness of the spark eroded and wire cut surfaces, the latter had much lower initial roughness values (Figure 7.3b). The wire cut and spark eroded surfaces had average surface roughnesses of 0.4 µm and 1.6 µm. The surface roughness decreased with each polishing treatment and converged to 0.09 µm (wire cut) and 0.14 µm (spark eroded). The average roughness of the mechanically polished reference samples B1 to B3 after the mechanical polishing was as followed: 0.15 µm (B1), 0.07 µm (B2), and 0.08 µm (B3). The average material removal during the mechanical polishing was 35 µm for B1 and B3 and 51 µm for B2.

101

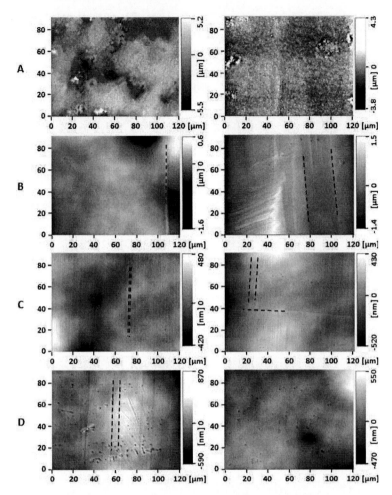

Figure 7.2: Micrographs of sample A5 taken with an optical profilometer. Spark eroded and wire cut surfaces are shown in the left and right columns, respectively. The four micrographs in each column are representing the surfaces (A) initially before the deformation cycle 1, (B) before cycle 2 (after the first electropolishing treatment), (C) before cycle 3 (after second electropolishing treatment), and (D) before cycle 4 (after third electropolishing treatment). Each micrograph shows an area of 121 µm x 92 µm.

The deformation experiments were evaluated regarding the 0.2% yield stress σ_Y, and the stress σ_{95} at 95% of the total strain. The twinning stress, σ_{tw}, is defined here as the mean value of the stress within the defined strain range. The twinning stress was evaluated by graphically integrating the stress values. The difference between the lowest and highest applied stress within these two strain marks define the twinning stress range, $\Delta\sigma_{tw}$.

The dashed lines exemplarily indicate the direction of straight ridges and valleys appearing after the first electropolishing treatment. These mark the traces of twin boundaries.

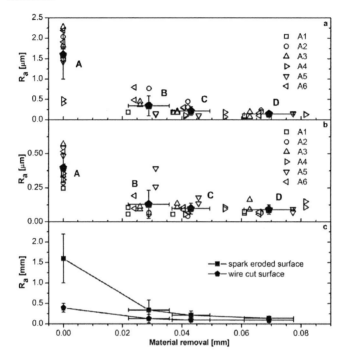

Figure 7.3: Roughness, Ra, for areas of 92 μm x 121 μm of the spark eroded (a) and wire cut (b) surfaces, and a comparison of both (c) after each polish treatment. Capital letters in (a) and (b) indicate the polishing state: (A) unpolished, (B) after 30 s, (C) after 60 s, (D) after 120 s. The solid pentagon is the average value. In (c), the squares and pentagons indicate the average roughness of the wire cut and spark eroded surfaces, respectively.

The stress-strain curves of the first deformation experiments under loading in z direction before the electropolishing and after each polishing treatment are giving in Figure 7.4. Table 7.3 provides a summary of the deformation tests for all samples. Specimens A1 to A4 show a drastic decrease of twinning stress after the first polishing treatment. The twinning stress after the second and third polishing treatment was constant or decreased slightly with each following polishing treatment. The stress-strain curves after the first, second, and third polishing treatment were very flat (i.e. they showed little hardening) and all twinning stresses were below 5 MPa with lowest values at 0.8 MPa (A1).

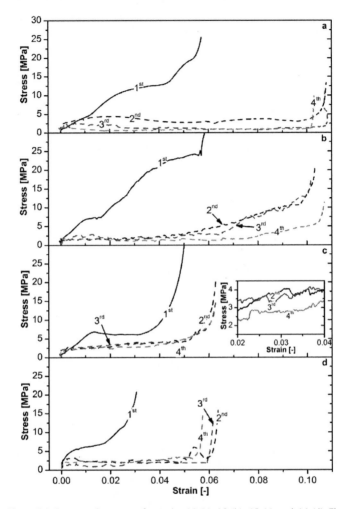

Figure 7.4: Stress-strain curves of samples A1 (a), A2 (b), A3 (c), and A4 (d). The solid lines indicate the loading curve before electropolishing, the dashed lines after electropolishing. In the inset in (c), all stress-strain curves are drawn with solid lines to emphasize the serrated flow.

Table 7.3: Average twinning stresses, σ_{tw}, and the twinning stress range, $\Delta\sigma_{tw}$, in square brackets, for samples A1 to A4 and B1 to B3 of test series A.

	Test series A, σ_{tw} [$\Delta\sigma_{tw}$], N/mm^2					
	1	**2**	**3**	**4**	**5**	**6**
A1	9.1 [17.8]	3.5 [1.9]	1.4 [1.8]	0.9 [1.7]		
A2	13.9 [21.8]	4.2 [9.1]	4.1 [10.1]	1.9 [4.7]		
A3	5.9 [7.2]	3.7 [4.9]	3.8 [4.7]	3.3 [5.0]		
A4	7.6 [9.4]	2.6 [3.2]	1.8 [1.8]	2.4 [1.5]	after mechanical polishing	
B1	13.0 [17.2]	3.8 [5.8]	2.8 [6.2]	2.1 [4.8]	1.8 [3.2]	1.3 [2.6]
B2	7.4 [12.7]	3.5 [6.3]	2.6 [5.8]	2.1 [4.8]	1.9 [2.6]	1.6 [2.7]
B3	14.6 [22.9]	4.9 [7.3]	2.8 [7.2]	3.1 [5.9]	1.9 [2.2]	1.4 [2.2]

The stress-strain curves of the control samples during test series A, B1 to B3, which were not electropolished, are shown in Figure 7.5. The first stress-strain curve of each unpolished sample had the same characteristics and twinning stress level (between 7.4 and 14.6 MPa, Table 7.3) as those of samples A1 to A6. After the first deformation test the twinning stresses dropped to values between 3.5 and 4.9 MPa. From the second deformation test on, the twinning stresses decreased only slightly during the subsequent deformation experiment. Finally, the twinning stresses of all B-samples were between 2.6 and 2.8 MPa. After the samples had been mechanically polished with a 9 μm and 6 μm polish, the average twinning stresses dropped further to 1.8 and 1.9 MPa. The twinning stress remained low over the entire strain range where twinning occurred before it sharply increased when twinning saturated. As a result of the nearly constant twinning stress of the mechanically polished sample, the twinning stress range decreased by half (indicated as error bar in Figure 7.6) from between 5.5 ± 0.4 MPa to 2.7 ± 0.5 MPa. This is in contrast with the unpolished samples, for which the stress levels increased constantly with increasing strain, and transitioned slowly to the elastic region.

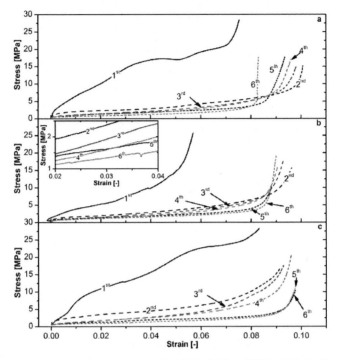

Figure 7.5: Stress-strain curves of samples B1 (a), B2 (b), and B3 (c). While loading 1 through 4 were performed with unpolished "as-cut" surfaces, loading 5 and 6 were performed with mechanically polished surfaces. The inset (b) illustrates the smoothness of the stress strain curves as compared to the serrated flow shown in the inset of Figure 7.4c.

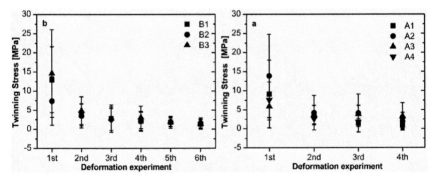

Figure 7.6: Twinning stresses, σ_{tw}, of the electropolished samples (a) and the control samples (b) during test series A. The bars indicate the twinning stress range, $\Delta\sigma_{tw}$ (not the experimental error).

The results of the deformation experiments of **test series B** are given in Table 7.4 and in Figure 7.7 (sample A2) and Figure 7.8 (sample A3). During the initial 2-dimensional deformation cycles the average twinning stress, σ_{tw}, of sample A2 decreased from 12.3 N/mm^2 (beginning of 1st cycle) to 5.2 N/mm^2 (beginning of 3rd cycle). Additionally, the twinning stress range, $\Delta\sigma_{tw}$, decreased from 12.2 N/mm^2 to 5.2 N/mm^2. Interestingly, the 10% mechanically induced strain which was formed for test series A was not reached in test series B. During the 2nd deformation cycle only 9.2% strain were reached. After the first 3-dimensional deformation cycle, σ_{tw} increased to 6.4 N/mm^2 and remained close to this value during the 3-dimensional training. After another 2-dimensional deformation cycle, σ_{tw} increased significantly to 10.5 N/mm^2 and with it the mechanically induced strain, which reached 12%. The twinning stress σ_{tw} decreased again during the 8th deformation cycle, and so did the mechanically induced strain. Interestingly, during this compression cycle in z direction, the dimension parallel to the y axis of the sample did not increased but the dimensions parallel to x axis of the sample. Therefore, the deformation direction was changed and an x-y twinning system established. This resulted in a significant increase of mechanically induced strain to 14.5% and 15% during deformation cycle 9 and 10. During these two deformation cycles, σ_{tw} stayed fairly constant at 8.0 and 8.4 N/mm^2 with twinning range of 4.8 and 4.9 N/mm^2, which were the lowest values observed in sample A2 for test series B.

Sample A3 (Figure 7.8) showed a much clearer twinning stress behavior than sample A2. While the mechanically induced strain during 2-dimensional deformation cycles was always below 6%, it increased to 7.2 during 3-dimensional deformation. σ_{tw} decreased during both 2-dimensional deformation cycle series (cycles 1-3 and 6-9) from 3.7 N/mm^2 to 2.9 N/mm^2 and from 6.5 N/mm^2 to 4.2 N/mm^2, respectively. During the 3-dimensional deformation cycles (4 and 5), σ_{tw} increased significantly to 9.3 N/mm^2 (cycle 4) and even 10.2 N/mm^2 (cycle 5).

Table 7.4: Average twinning stresses, σ_{tw}, and the twinning stress range, $\Delta\sigma_{tw}$, in square brackets, for samples A2 and A3 during test series B.

Test series B, σ_{tw} [$\Delta\sigma_{tw}$], N/mm^2				
1	**2**	**3**	**4**	**5**
A2 12.3 [12.2]	5.8 [7.7]	5.2 [6.2]	6.4 [13.8]	5.5 [8.4]
A3 3.7 [7.0]	3.3 [3.5]	2.9 [2.1]	9.3 [8.7]	10.2 [12.1]
6	**7**	**8**	**9**	**10**
A2 6.2 [10.2]	10.5 [19.3]	6.5 [11.8]	8.4 [4.9]	8.0 [4.8]
A3 6.5 [10.8]	4.7 [7.4]	4.2 [7.0]	4.2 [8.2]	-

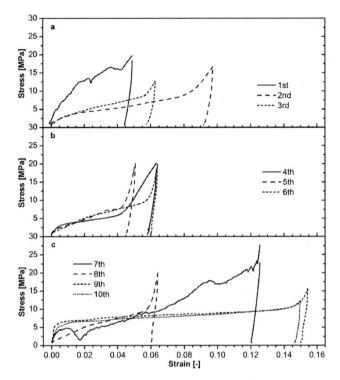

Figure 7.7: Stress-strain curves of sample A2 of compressive deformation in *z* direction in a 2-dimensional deformation, *z-y* (a), 3-dimensional deformation, *z-y-x* (b), and again 2-dimensional deformation mode, *z-x*, 7th deformation, and *x-y*, 8th and 9th deformation (c).

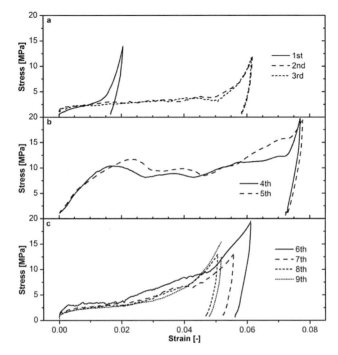

Figure 7.8: Stress-strain curves of sample A3 of compressive deformation in z direction in a 2-dimensional deformation, z-x (a), 3-dimensional deformation, z-y-x-y (b), and again 2-dimensional deformation mode, z-x (c).

During the 2-dimensional and 3-dimensional deformation cycles of sample A3, optical micrographs of the sample's surfaces were taken. During 2-dimensional compressive deformation, no traces of 45° twins, which indicate {110}-type twinning planes, were found on the z-y surfaces. During 3-dimensional compressive deformation, traces of 45° twins were indentified on the surfaces parallel to the z-y plane (see Figure 7.9) with the sample completely compressed.

Figure 7.9: Optical micrograph of surface parallel to *z-y* axes of sample A3 after (3-dimensional) deformation cycle 5.

When sample A1 and A3 were tested during **test series C** (see Figure 7.10), the average twinning stresses were 1.9 MPa and 3.7 MPa, respectively, in the electropolished condition (surface treatment 1). After the 9 μm mechanical polish (surface treatment 2), twinning stresses increased to 3.5 MPa and 4.2 MPa. A stronger change in twinning stress and stress-strain curve characteristics happened after the 1200 grit SiC grinding (surface treatment 3): twinning stresses increased to 4.8 MPa and 6.8 MPa with a smoothening of the stress-strain curve. Glass ball blasting (surface treatment 4) increased the twinning stresses even further to 7.6 MPa and 8.4 MPa. At the same time, the twinning stress ranges increased during these surface treatments as well from 1.4 MPa to 7.6 MPa for sample A1 and 4.5 MPa to 9.7 MPa for sample A3. An overview of this data is given in Table 7.3.

Table 7.5: Average twinning stresses, σ_{tw} and the twinning stress range, $\Delta\sigma_{tw}$ in square brackets, for samples A1 and A3 during test series C.

| | Test series C, σ_{tw} [$\Delta\sigma_{tw}$], N/mm^2 | | | |
	1	2	3	4
A1	1.9 [1.4]	3.5 [2.8]	4.8 [6.6]	7.6 [7.6]
A3	3.7 [4.5]	4.2 [5.9]	6.8 [5.9]	8.4 [9.7]

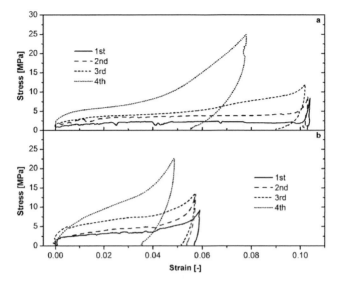

Figure 7.10: Stress-strain behavior of initially electropolished crystals after deforming the surface layer during test series C. (a) Stress-strain behavior of sample A1 (14 M martensite). Curve 1 shows the stress-strain behavior with electropolished surfaces, curve 2 with subsequently mechanically polished (9 μm diamond powder), curve 3 with subsequently grinded surfaces with 1200 grit SiC paper, and curve 4 with subsequently glass ball blasted surfaces. (b) shows the corresponding curves of sample A3 (10 M martensite).

Figure 7.11 and Figure 7.12 show micrographs taken of samples A1 and A3 after the loading cycles test series B. After these micrographs were taken, the next surface treatment was performed. Figure 7.11 shows the micrographs of sample A1 for the mechanically polished (a), mechanically grinded (b), and glass ball blasted (c) surfaces. For sample A3 (Figure 7.12), micrographs are shown for the electropolished (a), mechanically polished (b), mechanically grinded (c), and glass ball blasted (d) surfaces. While twins in the electropolished condition were rare and large (1 mm and larger), finer twins were appearing in the mechanically polished surfaces (0.2 mm). They became even finer (<0.1 mm, with some larger ones in sample A1) and more homogenously distributed.

The images shown in Figure 7.13 were taken with the optical profilometer and illustrate qualitatively how the surface roughness changed with each surface treatment during test series C. The initial surface, which was very smooth (a) was the result of the electropolishing treatments during test series A. After the sample was mechanically polished (b), the surface showed a few shallow polishing marks in form of long scratches over the surface. The depth and width of the scratches increased significantly after the mechanical grinding treatment (c). Glass ball blasting did not seem to have had a big effect, such as

111

dents on the surface, on the observed regions of 92 µm x 121 µm (in Figure 7.13 only section of 50 µm x 50 µm are shown). The relatively large size of the glass balls used during blasting, caused larger dents than the observed regions.

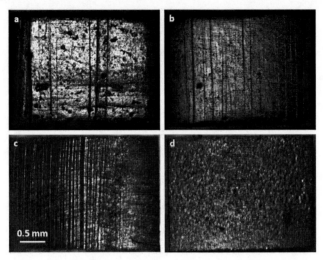

Figure 7.11: Optical micrographs of sample A1 (14 M martensite). (a) shows the micrographs of the electropolished, (b) of the mechanically polished (9 µm diamond powder), (c) subsequently grinded with 1200 grit SiC paper, (d) subsequently glass ball blasted surface.

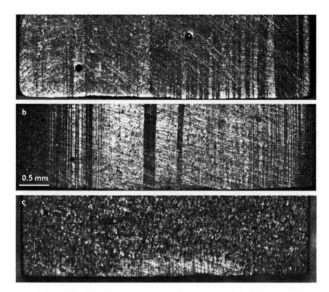

Figure 7.12: Optical micrographs of sample A3 (10 M martensite). (a) shows the micrographs of the mechanically polished (9 μm diamond powder), (b) subsequently grinded with 1200 grit SiC paper, (c) subsequently glass ball blasted surface.

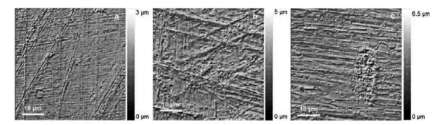

Figure 7.13: Micrographs of sample A3 taken with an optical profilometer. The three micrographs are representing the surfaces (a) after a mechanical polish with a 9 μm, (b) after a mechanical grinding treatment with a 1200 grit SiC paper, and (c) after the surface was glass ball blasted. Each micrograph shows a section of an optical profilometer scan of 50 μm x 50 μm.

During test series C, the average surface roughness, R_a, of sample A3 decreased slightly from 140 nm of the electropolished surface to 110 nm of the mechanically polished, probably since pitting holes were removed during mechanical polishing. With mechanical grinding, the average surface roughness increased to 620 nm. The surface roughness stayed constant after sand ball blasting. This might be an artifact caused by the small area, which was optically profiled. The results of the average surface roughness are given in Table 7.6. To characterize the extent and nature of the surface deformation, the full width half maximum

113

(FWHM) of diffraction reflections was analyzed. Using XRD, the FWHM of the 2θ peaks and rocking curves (ω scan) of {400} fundamental structure reflection was examined (Table 7.6) for surface treatments 2 to 5. For samples A1 and A3, the 2θ FWHM increased from surface treatment 2 to 4 from 0.56° to 0.61° and 0.36° to 0.81 for sample A1 and A3, respectively. After the chemical etching (surface treatment 5), the 2θ FWHM decreased again to 0.52° (A1) and 0.22° (A3). A3 further showed shift of the peak maximum of the $(004)_{14M}$ reflection between surface treatment 3 and 4. The modulation structure of the surface layer of A3 change from 10M to 14M after surface treatment 3 (appearance of the $(004)_{14M}$ reflection) and again after surface treatment 5 (disappearance of the $(004)_{14M}$ reflection). XRD 2θ scans of sample A1 and A3 are given in Figure 7.14. The change martensite structure was the reason why the ω scan was performed on sample A3 after surface treatment 2 and 5 for the $(400)_{10M}$ and after surface treatment 3 and 4 for the $(004)_{14M}$ reflection (see Figure 7.15b). The FWHM of the ω scan $(400)_{10M}$ reflection increased from surface treatment 2 to 5 and so did the FWHM of the $(004)_{14M}$ reflection from surface treatment 3 to 4. The FWHM of the ω scans of sample A1 (see Figure 7.14b) increased after surface treatment 3 and then decreased after the chemical etching (see Table 7.6).

Table 7.6: X-ray diffraction data of test series B after surface treatments 2-5. Full width half maximum (FWHM) is given for 2θ scans and ω scans (rocking curve) samples A1 and A3. The average surface roughness was only measured for sample A3.

| | | | **Sample A1** | | | |
Surface treatment		1	2	3	4	5
2θ FWHM	[°]	n/a	0.56	0.57	0.61	0.52
ω FWHM	[°]	n/a	0.46	0.35	6.69	2.84
			Sample A3			
Surface treatment		1	2	3	4	5
2θ FWHM	[°]	n/a	0.36	0.42	0.81	0.22
ω FWHM	[°]	n/a	1.11	1.86	4.44	7.60
at 2θ	[°]		62.4	67.8	67.8	62.4
R_a	[nm]	140±50	110±20	620±170	620±160	>50 µm

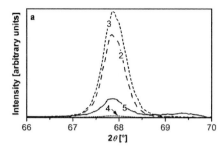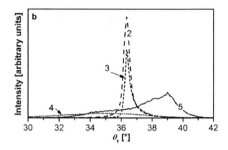

Figure 7.14: : X-ray Diffraction data of sample A1 with successive surface treatments 2-5 (a) 2θ scans of A1 with different surface treatments denoted by 2-5 as previously discussed, (b) rocking curves of A1 taken with a constant 2θ angle of 67.8°.

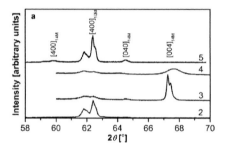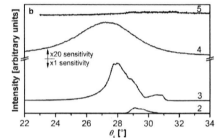

Figure 7.15: X-ray Diffraction data of sample A3 with successive surface treatments 2-5. (a) 2θ scans of A3 where numbers denote surface treatment. Curves (b) are rocking curves with curve numbers indicating the surface treatments. Rocking curves 2 and 5 were taken at a 2θ angle of 62.4° and rocking curves 3 and 4 were taken at a 2θ angle of 67.8°.

The neutron diffraction experiments of sample A3 yield lattice parameters $a = b = 6.03$ Å, $c = 5.59$ Å, and $\beta = 90.7°$ consistent with the monoclinic 10M phase. Sample B3 showed four fundamental reflections that can be associated with the {400} planes. The corresponding {100} d-spacings are 5.76 Å, 6.02 Å, and 6.48 Å, which are close to the lattice parameters of the orthorhombic 14M phase. The fourth fundamental reflection represents a d-spacing of 5.99 Å, which is approximately the a lattice parameter of the monoclinic 10M martensite. This indicates a mixture of 14M and 10M phases. Figure 7a shows a 001 section of the reciprocal space of sample A3 in a self-accommodated state. This diffraction pattern was obtained by rotating the sample around the axis perpendicular to the diffraction plane. The four equally spaced modulation reflections (subscript 'm' in Figure 7.16a) between the fundamental reflections (subscript 'f' in Figure 7.16a), e.g. between 040 and 220 reveal a 10M martensite structure. The 001 reciprocal space section of sample B3 in a self-accommodated state displays a 14M martensite structure based on the six modulation reflections (Figure 7.16b) between the fundamental reflections.

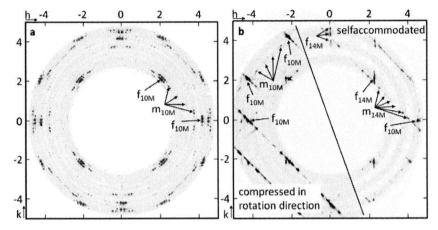

Figure 7.16: 001 sections of reciprocal space of the neutron diffraction data of sample A3 (a) and B3 (b) in the self-accommodated state. The number of modulation reflections indicate 10M martensite of sample A3 (4 modulation reflections, a) and 14M martensite of sample B3 (6 modulation reflections, b). Examples for modulation and fundamental reflections are marked with subscript 'm' and 'f'.

7.4. DISCUSSION

During **test series A**, the roughness of the spark eroded surface decreased from 1600 nm to just above 100 nm after three polishing treatments. For the wire cut surface the roughness decreased from 450 nm to just below 100 nm. About 80% of the reduction of the average surface roughness was reached during the first electropolishing treatment. While spark eroding is a very fast cutting technique, wire cutting resulted in a nearly four times lower average surface roughness. The initial roughness of the surface cut with the wire saw is similar to the roughness of the spark eroded surface after the first electropolishing treatment. This means that wire cutting reduces the number of sample preparation steps and eliminates the need of electropolishing if a surface roughness of 450 nm is sufficient.

For all samples, the average twinning stress dropped by about 50% after the first loading cycle and then by another 30% over the next two deformation tests. Deformation in two directions caused a selection of twin variants such that one single twinning mode was operational. With ongoing mechanical loading, the fraction of unfavorably oriented twins decreased. The mechanical behavior and twinning stresses were comparable for polished and unpolished samples. Thus, the removal of the thermally affected layer is not the controlling factor regarding twinning stress. The large reduction of twinning stress is caused mainly by mechanical softening. Nevertheless, polishing had an effect on the mechanical properties, namely at large strain. Here, hardening was reduced in polished samples as compared to unpolished samples. This effect is most evident when comparing the deformation curves of electropolished samples with those of unpolished samples as well as between the unpolished and mechanically polished states of samples B. A possible origin of the effect of surface roughness might be friction between sample and sample holder. Larger roughness may cause higher friction forces. This increased friction would suppress lateral expansion at large strain and therefore increase the load required for twinning.

The deformation tests also revealed a difference between electropolished and unpolished/mechanically polished samples. While the electropolished samples showed a serrated stress-strain curve after the first polishing treatment independent of the number of deformation experiments (inset of Figure 7.4c), the curves of the unpolished as well as later mechanically polished samples have smooth stress-strain curves (inset of Figure 7.5b). Figure 7.17 shows a schematic of the three surface treatment states, (a) spark eroded, (b) mechanically polished, and (c) electropolished. At the unpolished/spark eroded state (a), smooth deformation behavior may be caused by the increased surface roughness of the samples. An increased number of defects in the thermally influence surface layer (1) cause residual stresses and small radii surface features are resulting in stress concentrations, which act as finely dispersed pinning sites for twin boundaries. Thus, on very small scale, deformation is discontinuous (indicated by dashed lines (4) in Figure 7.17a and Figure 7.17b). However, these discontinuities are not resolved with the current testing method. The stress-strain curve appears as a smooth line.

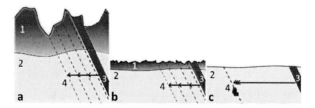

Figure 7.17: Schematic of surface layers of (a) the spark eroded, (b) the mechanically polished, and (c) the electropolished surface. (1) indicates a thermally or mechanically influenced layer with defects, (2) the not influenced bulk material, (3) a twin with preferred orientation relative to an applied stress during deformation experiment, and (4, dashed lines) positions at which twin boundaries arrest during deformation experiment. In (c), the twin boundary motion comes to a hold at an internal obstacle, while in (a) and (b), twin boundary motion stops due to surface irregularities.

The high friction between the sample and clamp may further affect the deformation behavior. Sudden expansion of the sample parallel to the surface might be constraint with increasing surface roughness. The expansion takes place in small, incremental steps at different positions throughout the sample wherever friction is locally lower than average (due to the uneven, unpolished surface). After several deformation experiments in the unpolished condition, a dual domain twin structure is established with fine traces of variants as described in [110,112]. Even in the deformed state these traces never completely vanish. Mechanical polishing reduces the surface roughness (Figure 7.17b). This reduces internal stresses and friction due to surface features. This reduces internal stresses and friction due to surface features. Thus, the twinning stress decreases. Furthermore, the stress at higher strains is reduced because of the reduced friction. This indicates that the statistical strength of obstacles in the sample is decreasing. Mechanical polishing partially removes defects caused by spark eroding, but also introduces new but smaller defects (Figure 7.17b, 1). The same arguments apply as for the unpolished sample. Therefore, the stress–strain curves are still smooth. A similar effect has been found in mechanically polished silicon wafers, where residual stresses due to defects on the surface caused by mechanical polishing, reached up to 5 µm deep into the silicon wafer [113]. After a short electropolishing treatment, the damaged surface layer was removed, and the residual stresses were reduced to nearly zero.

Electropolishing completely removes all surface defects (Figure 7.17c) and reduces residual stresses and stress concentrations at the surface. Thus, twin boundaries (4) can move very quickly over extended distances. The twin boundaries may be blocked intermittently at obstacles which are coarsely dispersed within the bulk of the crystal. This causes a discontinuous flow at a larger scale, which can be resolved. Therefore, the discontinuous flow appears as serrations in the stress-strain curves.

Another aspect that might have partially influenced the change of the stress-strain curves is the changed friction between the sample and clamp surfaces of the deformation test. In the unpolished, electropolished, and mechanically polished states of the samples, different surface roughnesses lead to different frictions. Sudden expansion of the samples parallel to the surface due to the twinning in large volume portions of the sample is constraint due to the high friction in unpolished samples. The expansion takes place in small, incremental steps at different positions throughout the sample wherever friction is locally lower than average (due to the uneven, unpolished surface). When the samples were polished, friction was reduced and larger motions due to the expansions perpendicular to the compression direction (parallel to the clamp surface) are possible.

During **test series B**, sample A2 and A3 were compressively deformed along two axes (2-dimenional deformation) and three axes (3-dimensional deformation). The mechanically induced strains of deformation tests of sample A2 revealed that there was probably a martensite phase change taking place from a 14M martensite to a NM martensite. This complicates the analysis of the twinning stress data. Nevertheless, during the 2-dimensional deformation test, a training effect which results in a decrease of the average twinning stress is visible. When compressively deformed in three dimensions, the average twinning stress did not decrease. In fact the twinning stress increased from the previous 2-dimensional deformation cycle. The change in martensite phase might have taken place during the 7th deformation cycle, when sample A2 was loaded with 28 MPa, 40% above the previously applied stress. Intermartensitic phase transformations especially from 14M to NM were described above in section 3.4.2.3 and have been observed in [56-58].

Sample A3 showed during 2-dimensional deformation decreasing twinning stress, and increasing twinning stress after the 3-dimensional deformation. Figure 7.18 shows an optical micrograph of the surface parallel to the z-y axes of the sample after the 3-dimensional deformation cycle z-y-x. Traces of twin in horizontal, vertical, and diagonal direction indicate that the deformations in y and x direction did not create single variant states. Traces of the previous variant state are still present. These twins might hinder twin boundary motion and might be the cause for the increased twinning stress during 3-dimensional deformation.

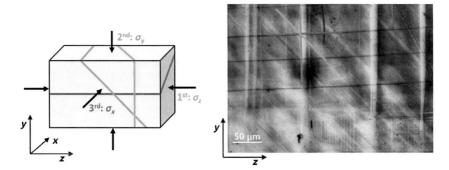

Figure 7.18: Schematic of the 3-dimensional deformation and optical micrograph of surface parallel to z-y axes of sample A3 after (3-dimensional) deformation cycle 5. The color coding indicates different twin directions and the deformation, from which they originate.

When increasing the surface roughness during **test series C**, the opposite effect takes place. The stress-strain curves shown in Figure 7.10 clearly reveal that surface deformations affect twin boundary motion. The slope of the stress-strain curves increases with increasing degree of surface deformation. Mechanical polishing creates surface layer defects which act as pinning sites for twin boundaries causing an increase of the twinning stress. The increasing slope of the curves, indicate a higher strength of the obstacles in the sample or at the sample surface (curves 2 in Figure 7.10a and Figure 7.10b). Stronger deformation (by grinding with 1200 grit SiC paper and further by glass ball blasting) causes a larger number of defects and possibly stronger defects. Therefore, the stress-strain curves are steeper than the curves for the crystal with less deformed surface layers after mechanical polishing. Figure 7.11 shows optical micrographs that correspond to treatments imposed on sample A1 shown in Figure 7.10a. Figure 7.12 shows the optical micrographs of figure A3 (without micrograph of electropolished condition). The micrographs reveal that twins become finer when the surface is more strongly deformed. This is consistent with the explanation provided above, i.e. twin boundaries are pinned by the surface layer defects. It is noticeable that the (non-linear) elastic strain increases with increasing degree of deformation of the surface layer. For the glass ball blasted 14M sample, the strain recovers elastically from 7.5 % to 5.5 % and for the 10 M sample from 4.7 % to 3.2 %, when the stress was released. Glass ball blasting was used here to produce the strongest deformation. However, less deformation can be produced by blasting at greater distance or by using smaller, and possibly also softer, particles instead of glass balls. This means that one can tailor the deformation to achieve desired twinning stresses for magnetically driven applications. Additionally, tailored surface layer deformations can be used to keep prior produced twin structures (i.e. twin thickness) during magneto-mechanical and mechanical cycling.

Table 7.6 shows that the FWHM of the 2θ scans of the 400 type reflections increased with increasing degree of surface layer deformation. This indicates a larger amount of elastic strain in the surface layer, which is reduced again when the surface layer is chemically etched away. Thus, the peak broadening and thinning confirmed that the surface treatments caused an increase of localized strains in the surface layer. The associated stress concentrations pin the twin boundaries. The FWHM values in columns 2-4 of Table 7.6, correspond to curves 2-4 in Figure 7.10 and the optical micrographs shown in Figure 7.11 and Figure 7.12. The FWHM of the rocking curves of the 400 type reflections also increase from the mechanically polished surface to the grinded surfaces and further to the surfaces deformed by glass ball blasting. The broadening of the diffraction peaks in the rocking curves shows that the damage was the largest in the surface that was glass ball blasted. This means that glass ball blasting caused the largest misorientations in the surface layer. These misorientations may play a role in pinning twin boundaries and in the elastic recovery seen in the stress-strain curves (Figure 7.10). The change of fundamental martensite peak intensities of sample A3 shown in Figure 7.15 indicates an increasing volume portion of 14M martensite at the surface the surface was deformed. This can be also explained with the intermartensitic phase transformation from the 10M to the 14M phase described above in section 3.4.2.3 [56-58]. After chemical etching, the 14M peak intensities were reduced, because the surface layer with increased 14M martensite was etched.

The stress-strain curves were additionally analyzed with regard to the possible martensite structure present in the samples based on the maximum mechanically induced strain. Due to strains of around 6%, samples A3 and A4 were identified as mostly 10M. Samples A1, A2, B1, B2, and B3 displayed a maximum strain of around 10% (samples A1 and A2 even 10.5%), which is typical for 14M martensites.

The two Neutron diffraction experiments (Figure 7.16) of one sample of each group (i.e. exhibiting 6% and 10% mechanically induced strain) confirmed the identification of the martensite modulation. Samples A3 and B3 have the 10M and 14M structure, which is in agreement with the maximum strain of 6% and 10%, respectively. Furthermore, the four identified {400} reflection of the diffraction pattern of sample B3 also indicate a mixture of 14M and 10M martensite. The volume fraction of the 10M martensite appeared to be small, since only 14M modulation reflection were visible in the 001 section of the reciprocal space (Figure 7.16b). This small 10M volume fraction though caused a slight reduction of the maximum strain from 10% to 8.8%. The neutron diffraction experiments revealed different martensite microstructures. Sample A3 showed a large number of reflections near 040 which indicates that many variants were present which may have been internally twinned. Since modulation reflections were present in both directions (e.g. between 040 and 220 as well as between the 040 and -220) with similar intensity. This indicates that variants with mutually perpendicular direction of modulation were present in comparable volume fractions. The diffraction pattern of sample B3 (14M) on the other hand indicates that different variants were present with different volume fractions. The modulation diffractions were visible between the 040 and 400 as well as between 240 and 420 but only partially between the 040 and -400 and between the -240 and -420 reflections. Additionally, the fundamental reflections do not show internal twins, which would be represented by peaks close to the main fundamental peak at the same 2θ, as seen in Figure 7.16a.

8. HIGH CYCLE MAGNETO-MECHANICAL TESTING

8.1. INTRODUCTION

In a magnetic field with changing directions MSMAs may undergo cyclic magnetoplastic deformation [13,35,69]. Experiments with a rotating magnetic field of constant field strength were introduced in 2002 [69]. During one field rotation, MSMA single crystals complete two magneto-mechanical cycles. Each cycle consists of one expansion and one contraction. For a given specimen, the MFIS and the threshold stress for twin boundary motion can be altered by training, which is described in detail in section 4.5. These trainings result in a predominant twin variant. Depending on training and structure, in a rotating magnetic field MFIS values of up to 10% were observed [13]. As defined earlier [22,85], training is said to be effective if the initial MFIS in rotating-field experiments is of the order of 1% or more. In contrast, training is said to be ineffective if the initial MFIS is about 0.1% or less. The crystal quality influences the effectiveness of training and the subsequent MFIS in a rotating magnetic field. Crystal quality is one reason for insufficient quantitative reproducibility of magneto-mechanical results.

When tested in a rotating magnetic field, samples which, after effective training, show initially large MFISs tend to fail by fracture after a moderate number of cycles. Samples with less effective training and reduced initial magnetic-field-induced strain tend to sustain very large numbers of magneto-mechanical cycles [22,85]. In Ref. [15], it was suggested that cracks nucleate where coarse twins interact and grow along twin boundaries. It is assumed that cracks nucleate if a critical number of twinning dislocations have formed a dislocation wall.

This chapter is aimed to further substantiate this assumption by testing the magneto-mechanical properties of a Ni-Mn-Ga sample G31B after high-cycle testing of 100 million MMC [22,85], and by examining the fracture surface of the previously studied sample after it was forcibly cracked. Additionally, the results of high cycle magneto-mechanical experiments with the DMMT (see section 4.4.2) of Berlin004 and Berlin005 samples, which received different trainings. are summarized. This part of chapter 8 was also published in different form in [38].

The results of the high cycle magneto-mechanical tests of the Berlin004 and Berlin005 samples were compared to DMMT results of G31B. The MFIS of G31B development over 100 million MMC is shown in Figure 8.1. G31SC100B was analyzed with x-ray tomography, optical microscopy, and magneto-mechanical tests to characterize cracks and the after-cycle performance. To determine the difference between deformation of a soft (σ_{tw}

< 0.8 MPa) and hard MSMA upon the application of a rotating magnetic field, sample Bilbao 001A was compared in the OMMD with Berlin005 05B.

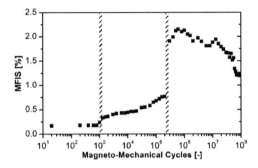

Figure 8.1: The development of the MFIS of Ni-Mn-Ga single crystal G31B. Long periods (regions left, center, right) during which MFIS changed slowly were separated from periods (gray vertical bars) over which MFIS increases rapidly. The MFIS increased from 0.2% during the first 1000 cycles to 2.15% at 0.6 million cycles and then decreased to 1.2% [22,85].

8.2. PROCEDURE OF EXPERIMENTS

In this study, Berlin004 and Berlin005 were tested in the DMMT (see section 4.4.2) for at least 1 million MMC (Berlin004) and for at least 0.5 million MMC (Berlin005) after they had received different trainings. Samples of Berlin004 column A were compressively deformed only along the samples' z direction. Samples of Berlin004 column E were additionally thermo-mechanically trained parallel to the dimension of the sample that expanded the most during the compressive deformation. Samples Berlin005 01B, 03C, and 05C were thermo-mechanically training and mechanically softened. Samples of single crystal ingot Berlin004 were positioned in the sample holder setup as described in section 4.4.2. During the course of this study, the sample holder design was slightly modified to uncouple the x and y component of the movement of the sample end that is attached to the sliding head (Figure 8.2) from the motion of the sliding head. The modification included the addition of a stabilizing decoupler (4) and a spring (5) at the sliding head providing a prestress of 0.1 to 0.15 MPa. This decoupling was performed to reduce the systematic error of the MFIS measurement due to motion of the sliding head not parallel to the z direction. The deformation of Berlin005 05C and B001B in a rotating magnetic field was analyzed in the OMMD before sample Berlin005 05C was tested in the DMMT. During the OMMD tests, both samples were constraint only on one side, i.e. glued to the sample holder but not to the sliding head. The applied magnetic field was with 1 T similar to the magnetic field of the DMMT.

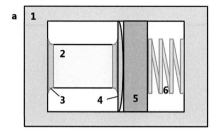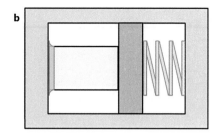

Figure 8.2: Schematics of DMMT sample holder with modifications for testing Berlin005: (a) sample (2) is glued to the sample holder/chamber (1) and to a stabilizing decoupler (4). The orange regions (3) indicate sample covered with glue. If the sample changes its shape in z direction, sliding head (5) is moving and redirects the sample deformation to the extensiometers. Spring (6) presses the sliding head against the decoupler and sample. Setup (b) is identical to (a) but without the decoupler.

After the sample G31B was exposed to a rotating magnetic field for 100 million MMC, x-ray tomography was applied to examine cracks, which were produced during cycling. Additionally, the twin structure and cracks on the surfaces were examined with optical microscopy. Then, the magneto-mechanical properties of the sample were tested in the SMMT (see section 4.4.1). While exposed to a constant magnetic bias field in y direction, the sample was mechanically loaded in z direction to a normal stress of up to 9 MPa, and is was subsequently unloaded. During the first test, the magnetic bias field strength was 2 T. The sample was unloaded in the field of 2 T to a minimum stress of less than 0.2 MPa. This minimum stress was required to hold the sample in place. Then the magnetic field was reduced to 0.15 T, and the sample was unloaded and loaded again. Unloading and loading experiments were repeated at orthogonal magnetic fields of 0.3 T, 0.35 T, 0.4 T, 0.5 T, 0.7 T, 1 T, and 2 T. In a final cycle, the sample was unloaded in a magnetic field of 2 T, and then loaded and unloaded without magnetic field. After magneto-mechanical testing, the crack development during these tests was studied using optical microscopy. Then, G31B was again magneto-mechanically actuated for approximately 20 cycles in a magnetic bias field of 0.9 T. The sample was finally broken forcibly and the fracture surface was examined with a scanning electron microscope (LEO 1430VP). In the result and discussion section, samples with a twinning stress below 1 MPa are referred to as "soft" MSMAs and sample with a twinning stress of 1MPa or higher as "hard" MSMAs.

8.3. RESULTS

The results of the magneto-mechanical cycling of samples Berlin004 column A and E are summarized in Figure 8.3a and b. Neither Berlin004 A (a) nor E (b), or neither mechanically trained (a) nor thermo-mechanically trained and mechanically trained (b) samples show MFIS above 0.4% Samples Berlin004 05A, 07A, and 11E do not show any change in MFIS over the magneto-mechanical cycling as well as an MFIS less than 0.01%. Samples Berlin004 09A, 11A, 03E and, 05E show a training effect during cycling but still less than 0.1% MFIS. Only samples Berlin004 03A, 04A, and 07E exceeded 0.1% MFIS.

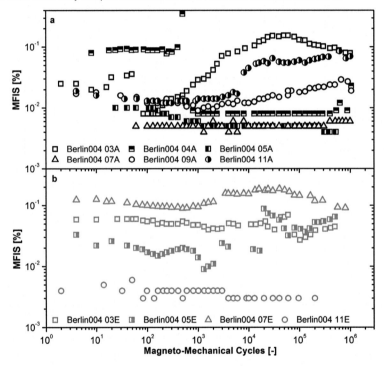

Figure 8.3: Results of high cycle magneto-mechanical tests of samples of column Berlin004 A (a) and E (b).

Before further high cycle tests were performed, sample Berlin005 05C and B001B were tested in the OMMD with only being glued on one side of the sample. Figure 8.4 shows the optical micrographs taken in the OMMD of samples Berlin005 05C (a) and B001B (b). Sample Berlin005 05C (a) started deforming very sudden when the angle between the sample's z axis and magnetic field direction reached approximately 60°. Two more sudden motions at field angles of 65° and 70° resulted then in the shapes shown for field angles of 65° and 90°. In contrast to this abrupt deformation behavior, sample B001B deformed very smoothly and over the entire volume of the sample. Deformation started shortly above 45°. Deformation states are given in Figure 8.4b for field angles of 0°, 50°, 55°, and 90°. During the deformation, the top surface - at 0° magnetic field orientation parallel to the bottom surface – stayed parallel for sample B001B at all magnetic field orientation. For sample Berlin005 05C the top surface was tilted by 4° at the two intermediate deformation states before it became parallel again after the deformation was complete at a field angle of 90°.

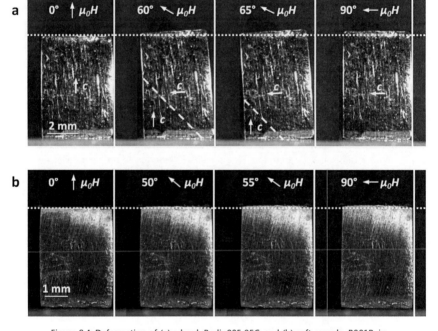

Figure 8.4: Deformation of (a) a hard, Berlin005 05C, and (b) soft sample, B001B, in a rotating magnetic field. In (a), one single twin boundary moves through the sample and is visualized by the dashed line. This single twin boundary divides the sample in two twin variants with differently aligned c axes. This results in the tilting of the upper part of the sample. The top surface is therefore not parallel to the bottom surface. In (b), twin boundaries are not visible but the sample is deforming in a more homogenous way and deforms constantly without tilting.

The high cycle magneto-mechanical test results of samples Berlin005 01B, 03C, and 05C are summarized in Figure 8.5a. The change of MFIS from constraint B to C of Berlin005 05C is detailed in Figure 8.5b.

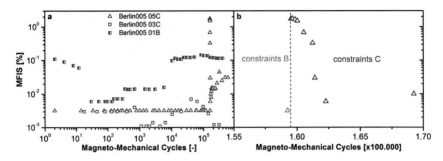

Figure 8.5: Results of high cycle magneto-mechanical tests of samples (a) Berlin005 01B, 03C, and 05C and (b) a detailed representation for sample Berlin005 05C. Constraints B (red) indicate tests with the sample glued on both end but without sample chamber lid, and constraint C (blue) only glued to the sample holder without decoupler.

Initially, sample Berlin005 05C was glued to the sample holder and decoupler (constraints A, see Figure 8.2a) during the high cycle tests. Even though the sample showed magnetically induced deformation in the OMMD, the MFIS in the high cycle test was only 0.003%. When removing the sample holder lid (Figure 4.18), constraints B, the MFIS did not change. When removing the decoupler (constraints C) resulted in an increase of MFIS by nearly three orders of magnitude to 2% strain. After just 1600 MMC, the MFIS decreased to less than 0.04% for the next 0.45 million MMC.

Sample G31B was examined for cracks with micro computer tomography (Figure 8.6) and optical microscopy after it had been cycled for 100 million cycles in a 0.97 T magnetic field. In the lower and upper quarter of the sample, all cracks were found to be parallel to each other, while in the center, there were also cracks perpendicular to each other. The distance between parallel cracks longer than 1 mm was 0.5 mm or more. All cracks formed an angle of approximately 45° with the lateral surfaces and, therefore, were parallel to {110}. The micro computer tomography slices revealed pores in the sub-100 µm range marked with arrows in Figure 8.6a. However, pores do not seem to be nucleation sites of cracks. Figure 8.6d shows an optical micrograph of one surface, which shows not only the cracks also visible in the x-ray tomography slices but also twins.

a   b   c   d

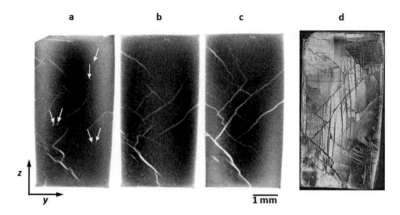

Figure 8.6: (a-c) X-ray micro tomography micrographs after 100 million MMC. Arrows in (a) indicate pores. Cracks are mostly 45° to the edges and therefore on {110} planes which are twinning planes. The grey shade is caused by beam hardening. (d) Optical micrograph of the surface close to slice (a). Cracks are visible over the entire surface, with a higher density in the center of the sample. Most cracks form an angle of 45° with the lateral surfaces, or are connecting cracks running at a 45° angle.

The recoverable strain in magneto-mechanical tests with an orthogonal magnetic bias field increased with increasing magnetic field (Figure 8.7a). Loading curves are solid, unloading curves dashed. The stress-strain curves are aligned at the recovered deformation at 0% strain. With increasing magnetic field, the stress needed to initiate twin boundary motion increased from 2 MPa at 0.35 T to above 3 MPa at 2 T. Also the stress at which strain recovery was started upon unloading increased with increasing magnetic field. The recoverable strain versus the applied magnetic field (Figure 8.7b) is below 0.3% in a magnetic field of up to 0.3 T. When the magnetic field was increased to 0.4 T, the recoverable strain rose quickly to 2.8%. At 0.7 T the recoverable strain was 5.7%, which was close to the maximum of 5.9% that was reached at 1 T and 2 T. After these tests, the sample was magneto-mechanically cycled several times, but no change in crack structure or length was observed.

Figure 8.8 shows the fracture surfaces after the sample was forcibly broken into two pieces. By comparing the micro computer tomography micrographs and the optical micrographs with the fracture surface, two different regions were identified: surface regions that were fractured during the 100 million cycling period and regions of the forced fracture surface. The fracture surface of the region that was cracked during the 100 million magneto-mechanical cycles (Figure 8.8a) is facetted with clear crystallographic features including small fracture steps with step heights of less than 1 μm (indicated by the arrows in Figure 8.8a). The fracture surfaces of the forced crack (Figure 8.8a) had a different fracture characteristic.

The surfaces of the forced crack showed additional microcracks that were parallel to the main fracture surface. These additional microcracks started at steps indicated by the arrows in Figure 8.8b.

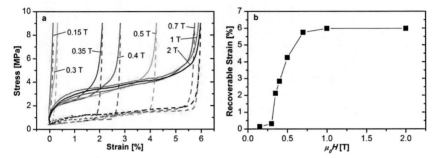

Figure 8.7: Magneto-mechanical properties of G31B after 100 million MMC. (a) Uniaxial compression stress versus strain curves at constant values of transverse magnetic field measured after 100 million magneto-mechanical cycles. At the fully compressed state (9 MPa in z direction), the magnetic field was changed. The lines with different shades of grey indicate different constant magnetic bias fields under loading (solid lines) and unloading (dashed lines). Recoverable strain for samples exposed to various magnetic field strengths.

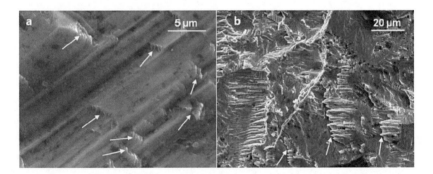

Figure 8.8: Scanning electron micrographs of the fractured surfaces. (a) shows the part of the fracture surface that was cracked before the sample was broken forcibly. The surface is micro-facetted. Between the layers of the crystal, there are very distinct ripple like steps (indicated by the arrows). (b) shows a region of the fractured surface resulting from the final forced fracture. The surface is very rough and other microcracks that seem to be parallel to the main fracture surface are visible (some are highlighted by arrows).

8.4. DISCUSSION

For the purpose of this discussion, samples with twinning stresses clearly above 2 MPa during a first mechanical deformation or at or above 1 MPa after mechanical softening are here defined as hard MSMAs. If they show a twinning stress below 1 MPa after mechanical softening, they are defined as soft MSMAs. Thus, all Berlin004 and 005 samples tested throughout this dissertation have to be considered hard MSMAs. While same samples of ingot Berlin005 (see chapter 6, Berlin005 03C and 05C) have at least after mechanical softening a twinning stress of below 2 MPa, Berlin004 samples have minimum twinning stresses just above 2 MPa (Berlin004 03C and 05C). Considering that the constraints of the gluing of the samples to the sample holder and sliding head has to be overcome additionally to the twinning stress, samples of ingot Berlin004 (glued on both ends) have very little potential to show significant MFIS in a rotating magnetic field. Additionally, samples Berlin004 have not been mechanically softened before mechanical cycling, so that they had even higher twinning stresses. As expected, the high cycle magneto-mechanical tests of sample Berlin004 A and E revealed that these hard samples have MFIS values of below 0.2%. The largest MFIS was observed at sample Berlin004 03A and 04A, which are from an ingot region with lower twinning stresses (see chapter 6). Interestingly, sample Berlin004 11A, which has probably a NM martensite structure based on results from chapter 6, shows a MFIS of up to 0.1%. This MFIS may be an artifact caused by a systematical error of the DMMT when the sample is glued to the sliding head. In that case, strain is recorded if the sliding head moves in x and y directions. This systematic error was eliminated by decoupling the sliding head from the sample. The weight of the extensiometer and the sliding head-ceramic rod assembly of the DMMT keep the sliding head on the bottom surface of the sample holder, so that it is more difficult to move vertically. Altogether, the measured very low MFIS of samples of ingot Berlin004 agrees with the high twinning stress shown in chapter 6, which is above the magneto-stress produced by the rotating magnetic field. Thus, twin boundaries are not expected to move.

This new sample holder design was employed during the high cycle tests of Berlin005 samples. In single crystal ingot Berlin005, samples 03C and 05C, characterized in chapter 6, showed the lowest twinning stress of 1.5 MPa (03C) and 1.0 MPa (05C) after repeated mechanical deformation (i.e. mechanical softening). Therefore, these and samples 01B, which also showed twinning stresses below 2 MPa of ingot Berlin005 have been chosen to be tested over 0.5 million MMC. Even though their twinning stress is lower than the twinning stress of samples of ingot Berlin004, they still fall into the category of hard MSMAs. Exemplarily depicted in Figure 8.4a, hard MSMAs show very sudden deformations with only one visible twin boundary moving through the sample. In opposite to the behavior of this hard MSMA, sample B001B, which is a soft MSMA, shows continues and smooth deformations. The difference in this deformation behavior is the reason, why sample Berlin005 05C showed significant MFIS after it was not glued anymore to the decoupler. Figure 8.9 shows schematics of the deformation of hard MSMAs glued on two sides (a) and on only one side (b). During the magnetically induced deformation of hard MSMAs, the samples tilt and bend as seen in Figure 8.4a. When glued on both sides, this tilting motion is not possible because the sliding head or decoupler, can only move slightly in y direction depending on the available gap to the side walls of the sample holder. Thus only very small MFIS is possible if any at all. The second glued side also adds another volume fraction in the sample that cannot deform. This increases internal stresses that might hinder twin boundary motion at all. Even if some volume fraction that is not constraint (blue region indicated in Figure 8.9a) has lower twinning stresses than the internal magneto stress, the missing degree of freedom to tilt restricts large MFIS. The MFIS is at detection limit (Figure 8.5b, constraints B), which supports this assumption. If the sample is only constraint on one side (Figure 8.9b), the free side of the sample is able to tilt and a twin boundary is free to move through the sample.

For soft samples, the second constraint is not an obstruction for the tilting of the sample since the soft MSMA does not visibly tilt (Figure 8.4b). This was demonstrated by sample G31B, which was constraint on both ends of the sample, but still showed MFIS over 1%. Additionally, sample B001B showed no tilting when rotated in the OMMD. This further supports the assumption that soft MSMAs have a more homogenous deformation behavior and can adapt to constraints much easier than hard MSMAs.

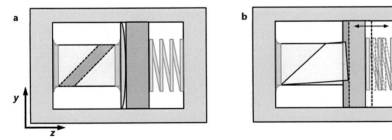

Figure 8.9: Schematic of the deformation of hard MSMAs in the decoupled DMMT sample holder setups. The hatching in (a) indicates the volume fraction in hard MSMA that is able to deform if mediator (4) would move in y direction. (b) shows how the sample deformations without decoupler and moves the sliding head.

Samples Berlin005 01B and 03C demonstrated lower MFIS than Berlin005 05C. This was expected due to higher twinning stress after mechanical softening. Even though Berlin005 05C had a MFIS of 2% after the decoupler was removed, the MFIS decreased to below 0.04% after just 1600 MMC. This reduction of the MFIS was not caused by cracks and is therefore only caused by hardening of the sample itself. This behavior was not observed before in samples that show MFIS above 1% [13,22,35,69,85]. Thus, hard samples are not only more difficult to be activated but also harden very easily.

The analysis of the x-ray tomography, scanning electron, and optical micrographs show that the reduction of MFIS of sample G31B from 2.15% to 1.2% is likely related to the cracks found in the micrographs. The micrographs further indicate that the cracks forming during the high-cycle magneto-mechanical testing grew along parallel crystallographic planes to a length of several millimeters with certain distinct micrometer size steps where the crack jumped from one crystallographic plane to a parallel plane.

The 45° angle between cracks and the {100} sample surface indicate that the cracks are formed on {110} planes, which are twinning planes. The cracks in the center of the sample are nearly perpendicular to each other. The intact regions between cracks span surface areas between 0.5 mm^2 and 2 mm^2. Magnetic-field-induced twins carrying a large magnetoplastic strain are coarse with a thickness in the order of 100 μm [11]. Orientation and size of the regions between the cracks suggest the presence of these large magnetic-field-induced twins. As described in Ref. [6], a larger number of dislocations pile up in walls between large interacting twins (unlike between thin, needle-like twins). With increasing number of dislocations in a wall, the stresses introduced at the dislocation walls eventually lead to the initiation and growth of cracks along the interface of the blocking twin [114,115]. As shown in Ref. [85], the initial twin thickness was below 10 μm. Owing to magneto-mechanical cycling, these thin twins merged by overcoming obstacles and formed larger twins. During the first 200,000 magneto-mechanical cycles, the twin thickness did apparently not reach the critical thickness [35] above which dislocation walls are large enough to induce critical stresses to initiate crack growth. Then, above 200,000 cycles, larger twins were formed and twin boundaries were able to sweep through a larger portion of the specimen, thus generating a larger MFIS (Figure 8.1). These larger twins also form larger dislocation walls inducing cracks that grow slowly until they are intercepted by other cracks growing perpendicular to them. This can be seen in Figure 8.6d where most cracks are terminating at intercepting perpendicular cracks.

By compressing the sample in the static magneto-mechanical test (where the sample is not glued) after the magneto-mechanical cycling, all twins – even those twins formerly restrained by the glue in the dynamic magneto-mechanical test over 100 million cycles – were aligned with the c direction parallel to the direction of compression. The recovered strain was 5.9%. This is nearly 3 times larger than the maximum MFIS during the high-cycle test with a rotating field. The lower magnetic field of 0.97 T in the dynamic magneto-mechanical test does not account for this reduction because the recoverable strain reaches a plateau-like maximum above 0.7 T (Figure 8.7a). The lower value of maximum recovered strain during the high-cycle test was probably caused by the constraints imposed by gluing the sample to the holder [85]. The microcracks parallel to the fracture surface of the forced crack confirm that the volume that was not cracked during the high-cycle magneto-mechanical tests consists of small twins with thicknesses in the micrometer range and below (as also evident from the ripples marked by arrows in Figure 8.8a). The small twins created during inefficient training are restrained from merging during the high-cycle magneto-mechanical test. A twin microstructure with thin twins contains weaker stress concentrations than a coarse twin microstructure, i.e. the stress distribution is more homogeneous [85]. Thus, in agreement with the observations, no cracking is expected in the sample corners. At first glance, one would expect that a sample with an extended network of cracks as present in the sample studied here, should display a significant degradation of magneto-mechanical properties. This study shows, however, that the magneto-mechanical properties of this sample are comparable to those of similar samples without cracks. After mechanical compression, the sample showed excellent magneto-mechanical properties over at least 20 loading-unloading cycles in a bias magnetic field without decay of the recoverable strain and without change of crack size and crack number. Thus, the functionality of MSMAs is not susceptible to degradation due to cracks, as long as mechanical integrity is maintained. This damage tolerance of MSMAs originates in the low stress level needed for twin boundary motion and the equally low stress level generated through a magnetic field (i.e. the magnetostress). Damage occurs at sites of stress concentration. Consequently, development of high-performance devices with MSMA transducers requires managing stress distributions through design and control of the microstructure.

From this chapter of the dissertation, is can be concluded that samples of single crystal ingot Berlin004 and Berlin005 are hard MSMAs. These hard MSMAs need mechanical softening to be able to shown MFIS. Furthermore, hard MSMA must not be constraint in a way so that they cannot tilt during deformation. From the high cycle study of G31B it can be concluded that even after 100 million magneto-mechanical cycles an ineffectively trained sample containing cracks can still have magneto-mechanical properties comparable to those of samples without magneto-mechanical cycling and not containing cracks. The results support the assumption that cracks nucleate where coarse twins interact whereas a densely twinned microstructure is more fracture resistant.

9. GENERAL DISCUSSION AND OUTLOOK

In this dissertation, sample preparation parameters, e.g. surface treatments and surface deformation, and different training methods, e.g. 2-dimensional and 3-dimensional mechanical softening, were deconvoluted and their influence on magneto-mechanical properties systematically and quantitatively investigated on more than just one sample. Furthermore, the influence of constraints on hard and soft MSMAs during magneto-mechanical cycling in a rotating magnetic field was identified. Additionally, the influence of different training methods on the high cycle magneto-mechanical properties of Ni-Mn-Ga MSMAs was examined.

This dissertation also gives a detailed and comprehensive overview of the properties of MSMAs depending on their position within a grown single crystal ingot for two nominal compositions: $Ni_{49.0}Mn_{30.0}Ga_{21.0}$ (ingots Berlin004 and Berlin005) and $Ni_{49.7}Mn_{29.3}Ga_{21.0}$ (ingot Berlin054). All properties including Curie temperature, martensitic phase transformation temperatures, magnetic anisotropy, saturation magnetization, martensite structure, lattice parameters, twinning stress and complex magneto-mechanical behavior depend on the composition of each sample which changes continuously in growth direction. Due to chemical segregation, the Mn content increases constantly in growth direction. With increasing Mn content the martensitic transformation temperature increases and the Curie temperature decreases. Most importantly, the martensite structure changes in growth direction from 10M over 14M to NM martensite. It was also shown that upon mechanical deformation, the martensite structure might be changed from 10M to 14M and from 14M to NM. This is in agreement with results of Chernenko et al. [58], who described the NM structure as ground state of Ni-Mn-Ga MSMAs. Comparing the dependency of the identified martensite structures on the composition as found in this study with the findings of Lanska et al. [87] and Richard et al. [88] some agreements and also discrepancies were revealed. The discrepancies are probably due to different methods employed for composition characterization, sample production methods, surface conditions, and sample form (i.e. single crystalline vs. powder).

Chapter 6 shows that polishing of the rough surface resulting from spark erosion is not solely responsible for reducing the twinning stress. Additional mechanical training is required to reduce the twinning stress significantly. Polishing reduces the twinning stress at higher strain by approximately 50%, but at lower strains by only 20%. This can be for recovery in a magnetic field. Furthermore, it was shown that mechanical softening by repeated compressive deformation occurs for polished and unpolished samples. Training is most effective when the samples are mechanically polished or electropolished prior to repeated deformation experiments. Surface damage and residual stress localized in the surface layer hinder and even prevent twin boundaries from moving. The stronger the surface layer damage, the higher the twinning stress and the larger the slope of the stress-strain curve. In the samples with surfaces containing defects due to different surface treatments (e.g., spark erosion, laser cutting, glass ball blasting and abrasive wearing including grinding and mechanical polishing using abrasive particles), twins are fine and the stress-strain curve is smooth due to a large number of densely dispersed pinning sites. When the defective layer is removed, e.g., by electropolishing, the twinning stress decreases and the twins coarsen in repeated compression/straining experiments. Without defective surface layer, the twin boundaries are pinned mainly by the coarsely dispersed obstacles (defects) in the bulk of the sample. It is also shown, that when a defect free surface (e.g. electropolished) is deformed, e.g. by one of the methods mentioned above, the stress-strain curve of a so prepared samples becomes steeper and smoother due to increased number of pinning sites. The larger the surface deformation, the steeper the stress-strain curve. Finally it was shown, that treatments that cause surface damage prevent the coarsening of the twins. They also keep the twin structure fine, if the twin structure is refined prior to the surface treatment.

In chapter 8, the influence of thermo-mechanical training as well as mechanical softening on hard MSMAs was investigated. In hard MSMAs, thermo-mechanical training does not seem to have an influence on the MFIS in high cycle magneto-mechanical tests. If any training method, mechanical softening should be employed to reduce twinning stresses and increase the chance of MFIS in rotating field actuations. Only one twin boundary was observed in hard MSMAs. This twin boundary moves through the entire sample. Thus, even if twinning stresses are reduced to a sufficient level of below 2 MPa, hard MSMAs must not be constraint in a way that they cannot tilt during deformation. If constraints resulting from clamping or holding mechanisms hinder tilting, the twin boundary will not be able to move through the sample. In such a case, constraints suppress MFIS and deform the sample macroscopically. It is further shown in chapter 8 that cracks in MSMAs do not need to result in failure or significant degradation of magneto-mechanical properties. Furthermore, cracks can stabilize over a large amount of MMC and might even reduce stresses between active portions and constraint sample portions. It was also shown that x-ray tomography can be utilized as nondestructive characterization method to analyze cracks and pores in MSMA single crystals.

The conclusions of this study are:

- All Ni-Mn-Ga properties depend strongly on composition and therefore due to chemical segregation on the position within single crystal ingots.
- The change in nominal composition from $Ni_{49.0}Mn_{30.0}Ga_{21.0}$ (Berlin004 and Berlin005) to $Ni_{49.7}Mn_{29.3}Ga_{21.0}$ (Berlin054) resulted in a much higher yield of samples with twinning stresses of below 2 MPa and increased the content of soft MSMA from zero to approximately 80%.

- Repeated consecutive mechanical loading in two dimensions reduces twinning stresses of polished and unpolished samples significantly.
- Repeated consecutive mechanical loading in three dimensions does not decrease twinning stresses.
- Surface deformation leads to pinning of twin boundaries, smoother stress-strain behavior and might be used to design specific twin microstructures.
- While electropolishing reduces the twinning stress most efficiently, mechanical polishing might be employed to moderately decrease twinning stresses and create finer twins.
- Hard MSMAs need to be mechanically softened to show MFIS. MFIS is obtained only if the sample is constraint on only one side, so that twin boundary motion can tilt the sample.
- Hard MSMA can harden during cyclic magneto-mechanical actuation.
- Cracks in MSMA do not need to result in the failure of MSMAs and might stabilize over magneto-mechanical cycling.

In the future, new production methods which reduce the amount of defects in Ni-Mn-Ga single crystals and which control the chemical segregation might lead to reduced twinning stresses and more homogenous distribution of martensite structures within one single crystal ingot. This could increase the yield of samples with large MFIS and therefore this could reduce the price of currently rather expensive Ni-Mn-Ga samples.

Furthermore, thermo-magneto-mechanical training might be utilized to increase or initialize MFIS of hard MSMAs. The thermal aspect of this training method, might be used as another type of in-service training.

To produce high-performing MSMAs with low twinning stress and high MFIS, single crystals have to be grown that have very little impurities and therefore very few pinning sites for twin boundaries. Furthermore, the nominal composition of the single crystals has to be chosen, so that their martensitic phase transformation temperature is close to operation temperature and their Curie temperature above that. And finally, at the moment 10M martensites are easier to train, but 14M martensites with MFIS of up to 10% can be an interesting alternative.

10. APPENDICES

10.1. APPENDIX A: LISTS

10.1.1. LIST OF ABBREVIATIONS

Abbreviation	Description
10M	10 layer monoclinic
14M	14 layer monoclinic
AFM	Atomic Force Microscope
BENSC	Berlin Neutron Scattering Centre
BSU	Boise State University
DFG	German Research Foundation
DMMT	Dynamic magneto-mechanical test
DSC	Differential Scanning Calometry
E2	Instrument at beamline 2 in the Experimental hall of BENSC
E3	Instrument at beamline 3 in the Experimental hall of BENSC
EDM	Electrical Discharge Machining
EDS	Electron Dispersive X-ray Spectrometry
EDX	See EDS
ETH	Eidgenössische Technische Hochschule
FWHM	Full width half maximum
HZB	Helmholtz Centre Berlin for Materials and Energy
ICP-AES	Inductive Coupled Plasma Atomic Emission Spectroscopy (same as ICP-OES)
ICP-OES	Inductive Coupled Plasma Optical Emission Spectroscopy
MFIS	Magnetic field-induced strain
MFM	Magnetic Force Microscope
MMMT	Manual Magneto-Mechanical Test
MMT	Manual Mechanical Test
MSMA	Magnetic shape-memory alloy
NM	Nonmodulated
OMMD	Optical Magneto-Mechanical Device
PMT	Premartensitic Phase Transformation
PMT	Intermartensitic Phase Transformation
PSI	Phase Shifting Interferometry
SE	Secondary Electrons
SEM	Scanning Electron Microscope
SLARE	SlAg Remelting and Encapsulating single crystal growth method
SMA	Shape memory alloy
SMMT	Static magneto-mechanical test
SPP1239	German Research Foundation Priority Program 1239
VSI	Vertical Scanning Interferometry
VSM	Vibrating Sample Magnetometer

10.1.2. LIST OF SYMBOLS

Symbols	Units	Description
μ	N/A^2	Permeability
μ_0	N/A^2	Magnetic constant, permeability of a vacuum
μ^i	μ^b	Magnetic Moment of Elements of Unit Cell
μ^i	μ^b	magnetic moment of elements of unit cell
μ_r	-	Relative permeability
μ^{tot}	μ^b	Total Magnetic Moment of Unit Cell
μ^{tot}	μ^b	Total magnetic moment of unit cell
2θ	°	Angle between incoming and diffracting beam
a^*, b^*, c^*	m^1	Reciprocal lattice parameter
$a_{10,x}$	Å	Lattice parameter a of 10M phase with x depending on unit cell
$a_{14,x}$	Å	Lattice parameter a of 14M phase with x depending on unit cell
A_f	K or °C	Austenitic finish temperature
a_{NM}	Å	Lattice parameter a of NM phase
A_p	K or °C	Austenitic peak temperature
A_s	K or °C	Austenitic start temperature
B	T	Magnetic flux density
$b_{10,x}$	Å	Lattice parameter b of 10M phase with x depending on unit cell
$b_{14,x}$	Å	Lattice parameter b of 14M phase with x depending on unit cell
b_{NM}	Å	Lattice parameter b of NM phase
b_t	m	Burgers vector
C	-	Curie constant
$c_{10,x}$	Å	Lattice parameter c of 10M phase with x depending on unit cell
$c_{14,x}$	Å	Lattice parameter c of 14M phase with x depending on unit cell
C_{Ga}	at.-%	Concentration of Ga
C_{Mn}	at.-%	Concentration of Mn
C_{Ni}	at.-%	Concentration of Ni
c_{NM}	Å	Lattice parameter c of NM phase
d_{hkl}	Å	Distance between [hkl] planes
d_t	m	Step height
e/a	-	Valence electron density
F_M	N	Magnetic Force
$FWHM$	°	Full width half maximum
$F_{x/y}$	N	Force on the Sample during heat treatment
$G_{austenite}$	J	Gibbs free energy of austenite phase
$G_{martenite}$	J	Gibbs free energy of martensite phase
H	T	Magnetic field strength
H_{app}	T	Applied magnetic field strength
H_c	T	Coercive field strength
H_{eff}	T	Effective magnetic field strength

H_{sat}	T	Saturation field
$H_{xz/yz}$	T	Magnetic field rotating in x-z/y-z plane
K	J/m^3	Magnetic anisotropy constant
K_a	J/m^3	Magnetic anisotropy between axes a and c
K_b	J/m^3	Magnetic anisotropy between axes b and c
M	A/m	Magnetization
M_f	K or °C	Martensitic finish temperature
MFIS	%	Magnetic-field-induced strain
m_{length}	m	Length of dimension parallel of the magnetic field direction
M_m	Am	Magnetization, Total magnetic moment per unit volume
m_{mag}	Am^2	Magnetic moment
M_p	K or °C	Martensitic peak temperature
M_r	A/m	Remanence
M_s	K or °C	Martensitic start temperature
M_{sat}	A/m	Saturation magnetization
M_V	A/m	Magnetization, Total magnetic moment per unit volume
N	-	Number of windings of a coil
N_D	-	Demagnetization factor
q	-	Shape based factor in calculation of demagnetization factor
R_a	m	Average surface roughness
s	-	Twinning Shear
s	-	Twinning shear
s_{mag}	m	Square root of cross section perpendicular of the magnetic field direction
t	s	Time
T_c	K or °C	Curie temperature
$T_{c,c}$	K or °C	Curie temperature upon cooling
$T_{c,h}$	K or °C	Curie temperature upon heating
T_m	K or °C	Martensitic phase transformation temperature
T_N	K	Néel temperature
$u_{A/B}$	J/m^3	Magnetic density of twin A/B
U_{ind}	V	Induced voltage
V	m^3	Volume
$\beta_{10,x}$	°	Angle between a and c of 10M phase with x depending on unit cell
$\beta_{10,x}$	°	Angle between a and c of 14M phase with x depending on unit cell
γ	°	Angle between x-Axis and Magnetic Field
ΔG_{a-m}	J	Gibbs free energy difference between phases at transformation upon cooling
ΔG_c	J	Difference in Gibbs free energy, chemical energy term
ΔG_e	J	Difference in Gibbs free energy, elastic energy term
ΔG_{m-a}	J	Gibbs free energy difference between phases at transformation upon heating

ΔG_s	J	Difference in Gibbs free energy, surface tension energy term
Δu	J/m^3	Energy difference between u_A and u_B
Δx	m	Elongation in x-direction
Δy	m	Elongation in y-direction
Δz	m	Elongation in z-direction
$\Delta \sigma_{tw}$	N/mm^2	Twinning stress range
ε_{max}	-	Maximal magnetic-field-induced Strain
$\varepsilon_{plastic}$	-	Plastic deformation strain
$\Theta_{1/2}$	°	Angle between incoming (1)/diffracted (2) beam and lattice plane
$\theta_{A/B}$	°	Deviation between M and axis of easy magnetization in twin A/B
λ	Å	Wavelength
ρ_{pores}	mm^{-3}	Density of pores
σ	Pa	Stress
σ_{tw}	N/mm^2	Average twinning stress
$\sigma_{x/y/z}$	N/mm^2	Applied stress in x, y, z direction
τ_M	Pa	Magneto-stress
Φ_M		Magnetic flux
χ	-	Magnetic Susceptibility
ω	°	Rotation of sample

10.1.3. LIST OF FIGURES

10.1.4. LIST OF TABLES

10.2. APPENDIX B: BIBLIOGRAPHY

[1]   Heusler, F., *Über magnetische Manganlegierungen*. Verhandlungen der Deutschen Physikalischen Gesellschaft e.v., **5**, 219(1903).

[2]   Heusler, F., *Über die Synthese ferromagnetischer Manganlegierungen*. Verhandlungen der Deutschen Physikalischen Gesellschaft e.v., **5**, 220(1903).

[3]   Starck, W. & Haupt, E., *Über die magnetischen Eigenschaften von eisenfreien Manganlegierungen*. Verhandlungen der Deutschen Physikalischen Gesellschaft e.v., **5**, 224(1903).

[4]   Rhyne, J.J., Foner, S., McNiff, E.J. & Doclo, R., *Rare earth metal single crystals; I. High-field properties of Dy, Er, Ho, Tb and Gd*. Journal of Applied Physics, **39**, 892-893

[5]   Liebermann, H.H. & Graham, C.D., *Magnetoplastic deformation of Dy crystals*. AIP Conference Proceedings, **29**, 598-599(1976).

[6]   Ullakko, K., Huang, J.K., Kantner, C., O'Handley, R.C. & Kokorin, V.V., *Large magnetic-field-induced strains in Ni2MnGa single crystals*. Applied Physics Letters, **69**, 1966-1968(1996).

[7]   Pons, J., Chernenko, V.A., Santamarta, R. & Cesari, E., *Crystal structure of martensitic phases in Ni-Mn-Ga shape memory alloys*. Acta Materialia, **48**, 3027-3038(2000).

[8]   Chernenko, V.A., Cesari, E., Pons, J. & Seguí, C., *Phase transformations in rapidly quenched Ni–Mn–Ga alloys*. Journal of Materials Research, **15**, 1496-1504(2000).

[9]   Bennett, J.C., Hyatt, C.V., Gharghouri, M.A., Farrell, S., Robertson, M., Chen, J. & Pirge, G., *In situ transmission electron microscopy studies of directionally solidified Ni–Mn–Ga ferromagnetic shape memory alloys*. Materials Science and Engineering A, **378**, 409-414(2004).

[10]   Wang, W.H., Liu, Z.H., Shan, Z.W., Chen, J.L., Wu, G.H. & Zhan, W.S., *Effect of post-growth annealing and magnetic field on the two-way shape memory effect of Ni52Mn24Ga24 single crystals*. Journal of Physics D: Applied Physics, **35**, 492-496(2002).

[11]   Murray, S.J., Marioni, M., Allen, S.M., O'Handley, R.C. & Lograsso, T.A., *6% magnetic-field-induced strain by twin-boundary motion in ferromagnetic Ni–Mn–Ga*. Applied Physics Letters, **77**, 886-888(2000).

[12]   Sozinov, A., Likhachev, A.A., Lanska, N. & Ullakko, K., *Giant magnetic-field-induced strain in NiMnGa seven-layered martensitic phase*. Applied Physics Letters, **80**, 1746(2002).

[13]   Müllner, P., Chernenko, V.A. & Kostorz, G., *Large cyclic magnetic-field-induced deformation in orthorhombic (14M) Ni–Mn–Ga martensite*. Journal of Applied Physics, **95**, 1531(2004).

[14]   Müllner, P., Chernenko, V.A. & Kostorz, G., *Large magnetic-field-induced deformation and magneto-mechanical fatigue of ferromagnetic Ni–Mn–Ga martensites*. Materials Science and Engineering A, **387-389**, 965-968(2004).

[15]   Müllner, P., Chernenko, V.A., Mukherji, D. & Kostorz, G., *Cyclic magnetic-field-induced deformation and magneto-mechanical fatigue of Ni-Mn-Ga ferromagnetic martensites*. Mater. Res. Soc. Symp. Proc. Volume 785, **785**, D12.2.1-D12.2.6(2004).

[16]   German Research Foundation (DFG) Priority Program *SPP1239 www.magneticshape.de*

[17]   Chernenko, V.A., Hagler, M., Müllner, P., Kniazkyi, V.M., L'vov, V.A., Ohtsuka, M. &
         Besseghini, S., *Magnetic susceptibility of martensitic Ni–Mn–Ga film*. Journal of Applied
         Physics, **101**, 053909(2007).

[18]   Gaitzsch, U., Potschke, M., Roth, S., Rellinghaus, B. & Schultz, L., *Mechanical training of
         polycrystalline 7M Ni50Mn30Ga20 magnetic shape memory alloy*. Scripta Materialia,
         57, 493-495(2007).

[19]   Gaitzsch, U., Potschke, M., Roth, S., Mattern, N., Rellinghaus, B. & Schultz, L., *Structure
         formation in martensitic Ni50Mn30Ga20 MSM alloy*. Journal of Alloys and Compounds,
         443, 99-104(2007).

[20]   Potschke, M., Gaitzsch, U., Roth, S., Rellinghaus, B. & Schultz, L., *Preparation of melt
         textured Ni–Mn–Ga*. Journal of Magnetism and Magnetic Materials, **316**, 383-
         385(2007).

[21]   Gaitzsch, U., Roth, S., Rellinghaus, B. & Schultz, L., *Adjusting the crystal structure of
         NiMnGa shape memory ferromagnets*. Journal of Magnetism and Magnetic Materials,
         305, 275-277(2006).

[22]   Chmielus, M., *Training, Microstructure, and Magneto-mechanical Properties of Ni-Mn-
         Ga*. (2007).

[23]   Boonyongmaneerat, Y., Chmielus, M., Dunand, D.C. & Müllner, P., *Increasing
         Magnetoplasticity in Polycrystalline Ni-Mn-Ga by Reducing Internal Constraints
         through Porosity*. Physical Review Letters, **99**, 14-17(2007).

[24]   Chmielus, M., Zhang, X.X., Witherspoon, C., Dunand, D.C. & Müllner, P., *Giant
         magnetic-field-induced strains in polycrystalline Ni–Mn–Ga foams*. Nature Materials, 2-
         5(2009).doi:10.1038/NMAT2527

[25]   Müllner, P., Zhang, X., Boonyongmaneerat, Y., Witherspoon, C., Chmielus, M. &
         Dunand, D.C., *Recent Developments in Ni-Mn-Ga Foam Research*. Materials Science
         Forum, **635**, 119-124(2010).

[26]   Li, Y., Jiang, C., Liang, T., Ma, Y. & Xu, H., *Martensitic transformation and magnetization
         of Ni–Fe–Ga ferromagnetic shape memory alloys*. Scripta Materialia, **48**, 1255-
         1258(2003).

[27]   Han, Z.D., Wang, D.H., Zhang, C.L., Xuan, H.C., Zhang, J.R., Gu, B.X. & Du, Y.W., *The
         martensitic transformation and the magnetocaloric effect in Ni50-xMn38+xIn12 alloys*.
         Solid State Communications, **146**, 124-127(2008).

[28]   Liu, Z.H., Wang, H.Y., Yu, S.Y., Dai, X.F., Chen, J.L., Wu, G.H. & Liu, Y., *Phase equilibrium
         of ferromagnetic shape memory alloy Co39Ni33Al28*. Scripta Materialia, **54**, 1299-
         1304(2006).

[29]   Dai, X.F., Wang, H.Y., Liu, G.D., Wang, Y.G., Duan, X.F., Chen, J.L. & Wu, G.H., *Effect of
         heat treatment on the properties of Co50Ni20Ga30 ferromagnetic shape memory alloy
         ribbons*. Journal of Physics D: Applied Physics, **39**, 2886-2889(2006).

[30]   Rolfs, K., Chmielus, M., Wimpory, R.C., Mecklenburg, A., Müllner, P. & Schneider, R.,
         Double twinning in Ni–Mn–Ga–Co. Acta Materialia, **58**, 2646-2651(2010).

[31]   Guldbakke, J.M., Chmielus, M., Rolfs, K., Schneider, R., Müllner, P. & Raatz, A.,
         *Magnetic, mechanical and fatigue properties of a Ni45.4Mn29.1Ga21.6Fe3.9 single
         crystal*. Scripta Materialia, **62**, 875-878(2010).

[32]   Liu, Z.H., Chen, J.L., Hu, H.N., Zhang, M., Dai, X.F., Zhu, Z.Y., Liu, G.D., Wu, G.H., Meng, F.B. & Li, Y.X., *The influence of heat treatment on the magnetic and phase transformation properties of quaternary Heusler alloy Ni50Mn8Fe17Ga25 ribbons.* Scripta Materialia, **51**, 1011-1015(2004).

[33]   Wu, G.H., Wang, W.H., Chen, J.L., Ao, L., Liu, Z.H., Zhan, W.S., Liang, T. & Xu, H.B., *Magnetic properties and shape memory of Fe-doped Ni[sub 52]Mn[sub 24]Ga[sub 24] single crystals.* Applied Physics Letters, **80**, 634(2002).

[34]   Guo, S., Zhang, Y., Quan, B., Li, J., Qi, Y. & Wang, X., *The effect of doped elements on the martensitic transformation in Ni–Mn–Ga magnetic shape memory alloy.* Smart Materials and Structures, **14**, S236-S238(2005).

[35]   Müllner, P., Mukherji, D., Aguirre, M., Erni, R. & Kostorz, G., *Micromechanics of magnetic-field-induced twin-boundary motion in Ni-Mn-Ga magnetic shape-memory alloys.* Solid-to-Solid Phase Transformation in Inorganic Materials 2005, **2**, 171-185(2005).

[36]   Müllner, P., Geleynse, A.S., Carpenter, D.R., Hagler, M.S. & Chmielus, M., *Modeling magnetoelasticity and magnetoplasticity with disconnections and disclinations.* Mater. Res. Soc. Symp. Proc. Volume 1050E, **1050**, (2008).

[37]   Chmielus, M., Carpenter, D.R., Geleynse, A.S., Hagler, M.S. & Schneider, R., *Numerical Simulation of Twin-Twin Interaction in Magnetic Shape-Memory Alloys.* Mater. Res. Soc. Symp. Proc. Volume 1090E, **1090**, (2008).

[38]   Chmielus, M., Chernenko, V.A., Hilger, A., Kostorz, G., Müllner, P. & Schneider, R., *Magneto-Mechanical Properties and Fracture of a Mechanically Constrained Ni-Mn-Ga Single Crystal after Extended Magnetic Cycling.* Proceedings of ICOMAT 2008, Santa Fe, NM, June 29–July 5, in press, (2010).

[39]   Buschow, K.H. & de Boer, F.R., *Physics of Magnetism and Magnetic Materials.* (Kluwer Academic Publisher: New York, 2004).

[40]   Craik, D.J. & Tebble, R.S., *Ferromagnetism and Ferromagnetic Domains.* Series of Monographs on Selected Topics in Solid State Physics, (1965).

[41]   Blakemore, J.S., *Solid State Physics.* (Cambridge University Press, Cambridge, United Kingdom: 1985).

[42]   Kasap, S.O., *Principle of Electronic Materials and Devices.* (McGraw-Hill, New York, USA: 2006).

[43]   Callister, W.D., *Materials Science and Engineering.* (John Wiley & Sons: New York, NY, USA, 2007).

[44]   Shackelford, J.F., *Introduction to materials science for engineers.* (Prentice-Hall Inc, Simon & Schuster/A Viacom Company: Upper Saddle River, NJ, USA, 1996).

[45]   Kaya, S., *On the Magnetisation of Single Crystals in Iron.* Science Report Tohoku University, **17**, 639(1928).

[46]   Honda, K. & Kaya, S., *On the Magnetisation of Single Crystals in Iron.* Science Report Tohoku University, **15**, 721(1926).

[47]   Gonzàlez-Comas, A., Obrado, E., Manosa, L., Planes, A., Chernenko, V.A., Hattink, B.J. & Labarta, A., *Premartensitic and martensitic phase transitions in ferromagnetic Ni2MnGa alloy.* Physical Review B, **60**, 7085-7090(1999).

[48]   Webster, P.J., Ziebeck, K.R., Town, S.L. & Peak, M.S., *Magnetic order and phase transformation in Ni2MnGa.* Philosophical Magazine B, **49**, 295-310(1984).

[49]   Block, T., *Neue Materialien für die Magnetoelektronik: Heusler- und Halb-Heusler-Phasen*. Johannes-Gutenberg-Universität Mainz, Mainz, Germany, (2002).

[50]   Ayuela, A., Enkovaara, J., Ullakko, K. & Nieminen, R.M., *Structural properties of magnetic Heusler alloys*. Journal of Physics: Condensed Matter, **11**, 2017-2026(1999).

[51]   Otsuka, K. & Wayman, C.M., *Shape Memory Materials*. (Cambridge University, Cambridge, United Kingdom: 1998).

[52]   Wu, S.K. & Yang, S.T., *Effect of composition on transformation temperatures of Ni-Mn-Ga shape memory alloys*. Materials Letters, **57**, 4291-4296(2003).

[53]   Chernenko, V.A., Cesari, E., Kokorin, V.V. & Vitenko, I.N., *The development of new ferromagnetic shape memory alloys in Ni-Mn-Ga system*. Scripta Metallurgica et Materialia, **33**, 1239-1244(1995).

[54]   Chernenko, V.A., Cesari, E., Pons, J. & Seguí, C., *Premartensitic phenomena and other phase transformations in Ni–Mn–Ga alloys studied by dynamical mechanical analysis and electron diffraction*. Physical Review B, **50**, 53-60(2002).

[55]   Seguí, C., Cesari, E., Pons, J. & Chernenko, V.A., *Internal friction behaviour of Ni–Mn–Ga*. Materials Science and Engineering A, **370**, 481-484(2004).

[56]   Chernenko, V.A., L'vov, V.A., Pons, J. & Cesari, E., *Superelasticity in high-temperature Ni–Mn–Ga alloys*. Journal of Applied Physics, **93**, 2394(2003).

[57]   Chernenko, V.A., Seguí, C., Cesari, E., Pons, J. & Kokorin, V.V., *Sequence of martensitic transformations in Ni-Mn-Ga alloys*. Physical Review B, **57**, 2659-2662(1998).

[58]   Seguí, C., Chernenko, V.A., Pons, J., Cesari, E., Khovailo, V. & Takagi, T., *Low temperature-induced intermartensitic phase transformations in Ni–Mn–Ga single crystal*. Acta Materialia, (2005).doi:10.1016/j.actamat.2005.07.018

[59]   O'Handley, R.C., *Model for strain and magnetization in magnetic shape-memory alloys*. Journal of Applied Physics, **83**, (1998).

[60]   Reinhold, M., Kiener, D., Knowlton, W.B., Dehm, G. & Müllner, P., *Deformation twinning in Ni-Mn-Ga micropillars with 10M martensite*. Journal of Applied Physics, **106**, 053906(2009).

[61]   Müllner, P., Clark, Z., Kenoyer, L., Knowlton, W.B. & Kostorz, G., *Nanomechanics and magnetic structure of orthorhombic Ni-Mn-Ga martensite*. Materials Science and Engineering A, **481**, 66-72(2008).

[62]   Overholser, R.W., Wuttig, M. & Neumann, D.A., *Chemical Ordering in Ni-Mn-Ga Heusler Alloys*. Scripta Materialia, **40**, 1095-1102(1999).

[63]   Schlagel, D.L., Wu, Y.L., Zhang, W. & Lograsso, T.A., *Chemical segregation during bulk single crystal preparation of Ni–Mn–Ga ferromagnetic shape memory alloys*. Journal of Alloys and Compounds, **312**, 77-85(2000).

[64]   Otsuka, K., Ohba, T., Tokonami, M. & Wayman, C.M., *New description of long period stacking order structures of martensites in beta-phase alloys*. Scripta Metallurgica et Materialia, **29**, 1359-1364(1993).

[65]   Pons, J., Santamarta, R., Chernenko, V.A. & Cesari, E., *Long-period martensitic structures of Ni-Mn-Ga alloys studied by high-resolution transmission electron microscopy*. Journal of Applied Physics, **97**, 083516(2005).

[66]   Jiang, C., Muhammad, Y., Deng, L., Wu, W. & Xu, H., *Composition dependence on the martensitic structures of the Mn-rich NiMnGa alloys*. Acta Materialia, **52**, 2779-2785(2004).

[67]   Kaufmann, S., Rößler, U.K., Heczko, O., Wuttig, M., Buschbeck, J., Schultz, L. & Fähler, S., *Adaptive Modulations of Martensites*. Physical Review Letters, **104**, 145702(2010).

[68]   Sozinov, A., Likhachev, A.A. & Ullakko, K., *Crystal Structures and Magnetic Anisotropy Properties of Ni–Mn–Ga Martensitic Phases With Giant Magnetic-Field-Induced Strain*. IEEE Transactions on Magnetics, **38**, 2814-2816(2002).

[69]   Müllner, P., Chernenko, V.A., Wollgarten, M. & Kostorz, G., *Large cyclic deformation of a Ni-Mn-Ga shape memory alloy induced by magnetic fields*. Journal of Applied Physics, **92**, 6708(2002).

[70]   Hull, D. & Bacon, D.J., *Introduction to Dislocations*. (Elsevier Butterworth-Heinemann: Oxford, United Kingdom, 2001).

[71]   Kostorz, G. & Müllner, P., *Magnetoplasticity*. Zeitschrift für Metallkunde, **96**, 703-709(2005).

[72]   Mecklenburg, A., Fiechter, S., Nabein, H. & Schneider, R., *patent DE102004018664A1*. (2005).

[73]   Wilke, K., *Kristallzüchtung*. (VEB Deutscher Verlag der Wissenschaften: 1973).

[74]   Brandon, D. & Kaplan, W.D., *Microstructural Characterization of Materials*. (John Wiley & Sons: Chichester, United Kingdom, 2003).

[75]   Goodhew, P.J., Humphreys, J. & Beanland, R., *Electron Microscopy and Analysis*. (Taylor & Francis: London, United Kingdom, 2001).

[76]   Whan, E.R., *ASM Handbook, Volume 10, Materials Characterization*. (ASM International: 1998).

[77]   Speyer, R.F., *Thermal Analysis of Materials*. (Marcel Dekker Inc.: New York, USA, 1994).

[78]   Wunderlich, B., *Thermal Analysis of Polymeric Materials*. (Springer: Berlin, Heidelberg, New York, 2005).

[79]   Baruchel, J., Buffiere, J., Maire, E., Merle, P. & Peix, G., *X-Ray Tomography in Material Science*. (Hermes Science Publication: 2000).

[80]   Olszak, A.G. & Schmit, J., *Interferometry: Technology and Application*. (Veeco Metrology Group, AN47 6/01: 2010).

[81]   Creath, K., *Temporal Phase Measurement Methods*. Interferogram Analysis, 94-140(1993).

[82]   Larkin, K.G., *Efficient nonlinear algorithm for envelope detection in white light interferometry*. Journal of the Optical Society of America A, **13**, 832-843(1996).

[83]   Wimpory, R.C., Mikula, P., Saroun, J., Poeste, T., Hofmann, M. & Schneider, R., *Efficiency Boost of the Materials Science Diffractometer E3 at BENSC: One Order of Magnitude Due to a Horizontally and Vertically Focusing Monochromator*. Neutron News, **19**, 16-19(2008).

[84]   Müllner, P., Chernenko, V.A. & Kostorz, G., *Stress-induced twin rearrangement resulting in change of magnetization in a Ni–Mn–Ga ferromagnetic martensite*. Scripta Materialia, **49**, 129-133(2003).

[85]   Chmielus, M., Chernenko, V.A., Knowlton, W.B., Kostorz, G. & Müllner, P., *Training, constraints, and high-cycle magneto-mechanical properties of Ni-Mn-Ga magnetic shape-memory alloys*. The European Physical Journal Special Topics, **158**, 79-85(2008).

[86]   Boonyongmaneerat, Y., Chmielus, M., Dunand, D.C. & Müllner, P., *Increasing Magnetoplasticity in Polycrystalline Ni-Mn-Ga by Reducing Internal Constraints through Porosity*. Physical Review Letters, **99**, 14-17(2007).

[87] Lanska, N., Söderberg, O., Sozinov, A., Ge, Y., Ullakko, K. & Lindroos, V.K., *Composition and temperature dependence of the crystal structure of Ni–Mn–Ga alloys.* Journal of Applied Physics, **95**, 8074(2004).

[88] Richard, M., Feuchtwanger, J., Schlagel, D., Lograsso, T., Allen, S. & Ohandley, R., *Crystal structure and transformation behavior of Ni–Mn–Ga martensites.* Scripta Materialia, **54**, 1797-1801(2006).

[89] Jin, X., Marioni, M., Bono, D., Allen, S.M., O'Handley, R.C. & Hsu, T.Y., *Empirical mapping of Ni–Mn–Ga properties with composition and valence electron concentration.* Journal of Applied Physics, **91**, 8222(2002).

[90] Khovaylo, V.V., Buchelnikov, V.D., Kainuma, R., Ohtsuka, M., Koledov, V.V., Shavrov, V.G., Takagi, T., Taskaev, S.V. & Vasiliev, A.N., *Phase transitions in Ni(2+x)Mn(1−x)Ga with a high Ni excess.* Physical Review B, **72**, 1-10(2005).

[91] Jiang, C., Liu, J., Wang, J., Xu, L. & Xu, H., *Solid–liquid interface morphology and crystal growth of NiMnGa magnetic shape memory alloys.* Acta Materialia, **53**, 1111-1120(2005).

[92] Sozinov, A., Söderberg, O., Koho, K., Sammi, T., Liu, X.W., Lanska, N. & Lindroos, V.K., *Effect of the selected alloying on Ni–Mn–Ga alloys.* Materials Science and Engineering A, **378**, 389-393(2004).

[93] Babita, I., Raja, M., Gopalan, R., Chandrasekaran, V. & Ram, S., *Phase transformation and magnetic properties in Ni–Mn–Ga Heusler alloys.* Journal of Alloys and Compounds, **432**, 23-29(2007).

[94] O'Handley, R.C., *Model for strain and magnetization in magnetic shape-memory alloys.* Journal of Applied Physics, **83**, 3263-3270(1998).

[95] Murray, S.J., Farinelli, M., Kantner, C., Huang, J.K., Allen, S.M. & O'Handley, R.C., *Field-induced strain under load in Ni–Mn–Ga magnetic shape memory materials.* Journal of Applied Physics, **83**, 21-23(1998).

[96] Vasil'ev, A.N., Bozhko, A.D., Khovailo, V.V., Dikshtein, I.E., Shavrov, V.G., Buchelnikov, V.D., Matsumoto, M., Suzuki, S., Takagi, T. & Tani, J., *Structural and magnetic phase transitions in shape-memory alloys Ni.* Physical Review B, **59**, 1113-1120(1999).

[97] Ullakko, K., Ezer, Y., Sozinov, A., Kimmel, G., Yakovenko, P. & Lindroos, V.K., *MAGNETIC-FIELD-INDUCED STRAINS IN POLYCRYSTALLINE Ni-Mn-Ga AT ROOM TEMPERATURE.* Scripta Materialia, **44**, 475- 480(2001).

[98] Popov, A.G., Belozerov, E.V., Sagaradze, V.V., Pecherkina, N.L., Kabanova, I.G., Gaviko, V.S. & Khrabrov, V.I., *Martensitic transformations and magnetic-field-induced strains in Ni50Mn50−x Gax alloys.* The Physics of Metals and Metallography, **102**, 140-148(2006).

[99] Seguí, C., Pons, J., Chernenko, V.A., Cesari, E., Ochin, P. & Portier, R., *Transformation and ageing behaviour of melt-spun Ni–Mn– Ga shape memory alloys.* Materials Science and Engineering A, **273-275**, 315 - 319(1999).

[100] Chopra, H.D., Ji, C. & Kokorin, V.V., *Magnetic-field-induced twin boundary motion in magnetic shape-memory alloys.* Physical Review B, **61**, 913-915(2000).

[101] Ge, Y., Heczko, O., Söderberg, O. & Hannula, S., *Magnetic domain evolution with applied field in a Ni–Mn–Ga magnetic shape memory alloy.* Scripta Materialia, **54**, 2155-2160(2006).

[102]   Zhou, Y., Jin, X., Xu, H., Kudryavtsev, Y.V., Lee, Y.P. & Rhee, J.Y., *Influence of structural transition on transport and optical properties of Ni[sub 2]MnGa alloy*. Journal of Applied Physics, **91**, 9894-9899(2002).

[103]   Söderberg, O., Ge, Y., Sozinov, A., Hannula, S. & Lindroos, V.K., *Recent breakthrough development of the magnetic shape memory effect in Ni–Mn–Ga alloys*. Smart Materials and Structures, **14**, S223-S235(2005).

[104]   Kudryavtsev, Y.V., Lee, Y.P. & Rhee, J.Y., *Structural and temperature dependence of the optical and magneto-optical properties of the Heusler Ni2MnGa alloy*. Physical Review B, **66**, 115114(2002).

[105]   Jääskeläinen, A., Ullakko, K. & Lindroos, V.K., *Magnetic field-induced strain and stress in a Ni-Mn-Ga alloy*. Journal de Physique IV France, **112**, 1005-1008(2003).

[106]   Cullity, B.D., *Introduction to magnetic materials*. 321(Addison-Wesley: Reading, MA, 1972).

[107]   Sozinov, A., Likhachev, A.A., Ullakko, K., Lanska, N. & Lindroos, K., *10% magnetic-field-induced strain in Ni-Mn-Ga seven-layered martensite*. Journal de Physique IV France, **112**, 955-958(2003).

[108]   Pasquale, M., Sasso, C.P., Besseghini, S., Villa, E. & Chernenko, V.A., *Temperature dependence of magnetically induced strain in single crystal samples of Ni–Mn–Ga*. Journal of Applied Physics, **91**, 7815(2002).

[109]   Müllner, P., Chernenko, V.A., Mukherji, D. & Kostorz, G., *Cyclic magnetic-field-induced deformation and magneto-mechanical fatigue of Ni-Mn-Ga ferromagnetic martensites*. Mat. Res. Soc. Symp. Proc. Vol. 785, **785**, D12.2.1-6(2004).

[110]   Straka, L., Heczko, O. & Hanninen, H., *Activation of magnetic shape memory effect in Ni–Mn–Ga alloys by mechanical and magnetic treatment*. Acta Materialia, **56**, 5492-5499(2008).

[111]   Chmielus, M., Rolfs, K., Wimpory, R., Reimers, W., Müllner, P. & Schneider, R., *Effects of surface roughness and training on the twinning stress of Ni–Mn–Ga single crystals*. Acta Materialia, **58**, 3952-3962(2010).

[112]   Straka, L., Novak, V., Landa, M. & Heczko, O., *Acoustic emission of Ni–Mn–Ga magnetic shape memory alloy in different straining modes*. Materials Science and Engineering A, **374**, 263-269(2004).

[113]   Dupke, R. & Reimers, W., *Evaluation of Near Surface Residual Stresses in Abrasive Machined Silicon Wafers*. Residual Stresses - European Conference on Residual Stresses, Nov 1992, 873-880(1993).

[114]   Müllner, P., Solenthaler, C., Uggowitzer, P.J. & Speidel, M.O., *Brittle fracture in austenitic steel*. Acta Materialia, **42**, 2211(1994).

[115]   Müllner, P., *On the ductile to brittle transition of austenitic steel*. Materials Science and Engineering A, **234-236**, (1997).

[116]   Sands, D.E., *Introduction to Crystallography*. (Dover Publication Inc.: New York, NY, USA, 1993).

[117]   Chernenko, V.A., Oikawa, K., Chmielus, M., Besseghini, S., Villa, E., Albertini, F., Righi, L., Paoluzi, A., Müllner, P., Kainuma, R. & Ishida, K., *Properties of Co-alloyed Ni-Fe-Ga Ferromagnetic Shape Memory Alloys*. Journal of Materials Engineering and Performance, (2009).doi:10.1007/s11665-009-9425-7

[118] Chernenko, V.A., Chmielus, M., Müllner, P., Oikawa, K., Besseghini, S., Villa, E., Albertini, F., Righi, L., Paoluzi, a., Kainuma, R. & Ishida, K., *Properties of Co-alloyed Ni-Fe-Ga Ferromagnetic Shape Memory Alloys.* Journal of Materials Engineering and Performance, **18**, 548-553(2009).

[119] Carpenter, D.R., Chmielus, M., Rothenühler, A., Schneider, R. & Müllner, P., *Modeling magnetoelasticity and magnetoplasticity with disconnections and dislocations.* Special Topics, (2008).

10.4. APPENDIX C: OVERVIEWS

10.4.1. RESULTS SUMMARIES: BERLIN004

Table 10.1: Summary of all characterized sample properties of single crystal ingot Berlin004 samples 03C, 05C, 07C, 09C and 11C

Property	unit	Berlin004				
		03C	05C	07C	09C	11C
C_{Ni}	at.-%	50.4±0.4	50.3±0.3	50.0±0.2	49.3±0.3	49.0±0.4
C_{Mn}	at.-%	27.9±0.3	28.5±0.4	29.0±0.3	29.9±0.4	30.9±0.4
C_{Ga}	at.-%	21.7±0.6	21.2±0.3	21.0±0.2	20.8±05	20.1±0.3
e/a	-	7.64±0.08	7.66±0.06	7.66±0.04	7.65±0.07	7.66±0.07
DSC: T_M	°C	77±1	81±1	82±1	90±1	91±1
DSC: T_c	°C	103±1	101±1	99±1	94±1	n/a
VSM: T_M	°C	69±1	74±1	73±1	82±1	95±1
VSM: T_c	°C	100±1	97±1	94±1	92±1	101±1
Neutron diffraction — a	Å	5.97±0.3	6.02±0.3	6.19±0.3	5.47±0.3	5.47±0.3
Neutron diffraction — b	Å	-	5.81±0.3	5.79±0.3	-	-
Neutron diffraction — c	Å	5.57±0.3	5.49±0.3	5.47±0.3	6.55±0.3	6.61±0.3
Neutron diffraction — c/a	-	0.929±0.005	0.907±0.005	0.876±0.005	1.210±0.005	1.222±0.005
modulation reflections		10M	14M	14M	NM	NM
$\rho_{pores > 25\ \mu m}$	mm^{-3}	4.0	2.5	2.6	3.1	3.9
$\rho_{pores > 50\ \mu m}$	mm^{-3}	0.3	0.2	0.5	0.5	0.4
cracks	-	1	1	1	6	17
$\varepsilon_{plastic}$	%	6.1	10.1	14.4	13.8	4.7
σ_{tw}	MPa	2.4	2.1	7.9	10.7	13.5
$\Delta\sigma_{tw}$	MPa	1.4	2.2	7.6	10.2	21.1
martensite based on σ-ε		10M	14M	NM/14M	NM	NM
K_a	10^5 J/m^3	1.63±0.15	1.78±0.15	1.67±0.15	-2.20±0.2	-2.02±0.19
K_b	10^5 J/m^3	-	0.80±0.07	0.69±0.07	-	-
M_{sat}	10^5 A/m	4.96±0.2	4.56±0.2	4.35±0.2	4.15±0.2	3.59±0.2

10.4.2. RESULTS SUMMARIES: BERLIN005

Table 10.2: Summary of all characterized sample properties of single crystal ingot Berlin005 samples 01C, 03C, 05C, 07C, and 09C.

		Berlin005				
Property	Unit	01C	03C	05C	07C	09C
C_{Ni}	at.-%	50.7±0.4	50.6±0.3	50.3±0.3	50.0±0.4	49.8±0.3
C_{Mn}	at.-%	27.8±0.2	28.0±0.3	28.5±0.3	28.9±0.4	29.6±0.3
C_{Ga}	at.-%	21.6±0.4	21.5±0.4	21.2±0.3	21.1±04	20.6±0.5
e/a	-	7.66±0.06	7.66±0.05	7.67±0.04	7.68±0.05	7.68±0.06
DSC: T_M	°C	83±1	85±1	88±1	90±1	97±1
DSC: T_c	°C	101±1	101±1	101±1	100±1	not meas.
VSM: T_M	°C	78±1	79±1	82±1	84±1	88±1
VSM: T_c	°C	98±1	98±1	97±1	94±1	96±1
neutron diffraction a	Å	5.91±0.3	5.95±0.3	6.02±0.3	6.21±0.3	$5.46_{NM}/6.22_{14m}$
b	Å	n/a	n/a	not identified	5.82±0.3	5.79_{14M}
c	Å	5.53±0.3	5.51±0.3	5.49±0.3	5.46±0.3	$6.58_{NM}/5.46_{14M}$
c/a	-	0.932±0.005	0.922±0.005	0.907±0.005	0.873±0.005	$1.22_{NM}/0.87_{14M}$
modulation reflections		10M	10M	14M	14M	NM
$\rho_{pores > 25\ \mu m}$	mm^{-3}	3.1	2.5	2.2	2.3	3.0
$\rho_{pores > 50\ \mu m}$	mm^{-3}	0.3	0.2	0.1	0.2	0.2
cracks	-	5	0	1	0	0
$\varepsilon_{plastic}$	%	4.6	10.3	10.4	16.1	9.7
σ_{tw}	MPa	9.7	1.5	1.0	8.2	14.0
$\Delta\sigma_{tw}$	MPa	9.7	1.8	1.2	7.4	23.1
martensite based on σ-ε		10M	14M	14M	NM	NM
K_a	10^5 J/m^3	1.52±0.14	1.45±0.13	1.69±0.15	-2.48±0.21	-2.22±0.2
K_b	10^5 J/m^3	1.52±0.14	1.45±0.13	0.71±0.06	-2.48±0.21	-2.22±0.2
M_{sat}	10^5 A/m	4.74±0.2	4.73±0.2	4.59±0.2	4.47±0.2	4.03±0.2

10.4.3. RESULTS SUMMARIES: BERLIN054

Table 10.3: Summary of all characterized sample properties of single crystal ingot Berlin054 samples 05A, 03A, and 01A.

Property	Unit	Berlin054 05A	03A	01A
C_{Ni}	at.-%	50.7±0.4	50.4±0.4	49.2±0.5
C_{Mn}	at.-%	26.1±0.3	27.2±0.2	29.6±0.4
C_{Ga}	at.-%	23.1±0.3	22.5±0.4	21.9±0.5
e/a	-	7.59±0.07	7.62±0.07	7.62±0.09
DSC: T_M	°C	51±1	57±1	72±1
DSC: T_c	°C	100±1	102±1	93±1
VSM: T_M	°C	44±1	49±1	59±1
VSM: T_c	°C	104±1	103±1	94±1
a	Å	5.99±0.3	5.99±0.3	5.55±0.3
b	Å	-	-	-
c	Å	5.62±0.3	5.62±0.3	6.61±0.3
c/a	-	0.938±0.005	0.938±0.005	1.191±0.005
modulation reflections		10M	10M	NM
$\rho_{pores > 25\,\mu m}$	mm^{-3}	n/a	n/a	n/a
$\rho_{pores > 50\,\mu m}$	mm^{-3}	0.2	0.2	0.1
cracks	-	n/a	n/a	n/a
$\varepsilon_{plastic}$	%	6.0	6.3	15.7
σ_{tw}	MPa	0.8	1.1	6.8
$\Delta\sigma_{tw}$	MPa	0.3	1.1	6.3
martensite based on σ-ε		10M	10M	NM
K_a	10^5 J/m^3	1.94±0.17	1.72±0.16	-2.17±0.2
K_b	10^5 J/m^3	1.94±0.17	1.72±0.16	-2.17±0.2
M_{sat}	10^5 A/m	5.39±0.25	5.27±0.25	4.46±0.2

(neutron diffraction spans rows a, b, c, c/a, modulation reflections)

157

10.4.4. Comparison of Martensite Structures

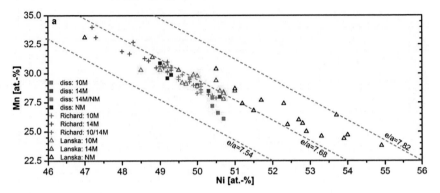

Figure 10.1: Comparison of in this work determined martensite phases (squares) with phases from Richard et al. [88] (crosses) and Lanska et al. [87] (triangles). The colors of the symbols indicate different martensite phases of mixed phases: 10M (red), 10M/14M (pink), 14M (blue), 14M/NM (green), and NM (black).

10.4.5. Overview of High Cycle Magneto-Mechanical Tests

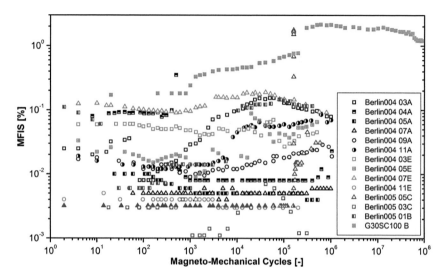

Figure 10.2: Comparison of all MFIS experiments of this study.

10.5. APPENDIX D: SUPPORTING MATERIAL

10.5.1. FUNDAMENTALS OF DIFFRACTION

10.5.1.1. RECIPROCAL SPACE

Diffraction experiments are based on the coherent elastic scattering of a wave by atoms. Diffraction can best be described in reciprocal space with the reciprocal lattice. Similar to the basis vectors of the crystal lattice $\boldsymbol{a},\boldsymbol{b},\boldsymbol{c}$, the reciprocal lattice is based on the translation vectors $\boldsymbol{a}^*,\boldsymbol{b}^*,\boldsymbol{c}^*$ in reciprocal space. In the case of an ideal infinite crystal diffracted intensity $I(\boldsymbol{H}^*)$ occurs only at the vectors

$$\boldsymbol{H}^* = h\boldsymbol{a}^* + k\boldsymbol{b}^* + l\boldsymbol{c}^* \tag{10.1}$$

Where h,k,l are the Miller indices of the plane. Thus, the reflection hkl form a lattice in reciprocal space. The direction of the reciprocal lattice vector $\boldsymbol{H}^*$ is normal to the (hkl) planes, while the length is reciprocal to the interplanar spacing d_{hkl}. The basis vectors of the real and reciprocal lattice satisfying the following conditions:

$$\boldsymbol{a}^* \cdot \boldsymbol{a} = \boldsymbol{b}^* \cdot \boldsymbol{b} = \boldsymbol{c}^* \cdot \boldsymbol{c} = 1 \text{ and } \boldsymbol{a}^* \times \boldsymbol{b} = \boldsymbol{a}^* \times \boldsymbol{c} = \boldsymbol{b}^* \times \boldsymbol{a} = ... = 0 \tag{10.2}$$

$$\text{and } \boldsymbol{a}^* = 2\pi \frac{\boldsymbol{b} \times \boldsymbol{c}}{\boldsymbol{a} \cdot (\boldsymbol{b} \times \boldsymbol{c})} \ , \ \boldsymbol{b}^* = 2\pi \frac{\boldsymbol{c} \times \boldsymbol{a}}{\boldsymbol{a} \cdot (\boldsymbol{c} \times \boldsymbol{a})} \ , \ \boldsymbol{c}^* = 2\pi \frac{\boldsymbol{a} \times \boldsymbol{b}}{\boldsymbol{a} \cdot (\boldsymbol{a} \times \boldsymbol{b})} \tag{10.3}$$

10.5.1.2. SCATTERING AT PERIODIC STRUCTURES, LAUE EQUATIONS

In 1912 Max von Laue at the University of Munich suggested that x-rays can diffract at periodic structures in crystals based on three assumptions [116]: (1) crystals are periodic, (2) x-rays are waves, (3) the wavelength of x-rays is in the same order of magnitude as the periodicity within crystals. Experiments by Friedrich and Knipping confirmed these assumptions. Von Laue created a mathematical framework for x-ray diffraction for 3-dimensional crystals. This framework is based on the diffraction of light by a fine grating (Figure 10.3).

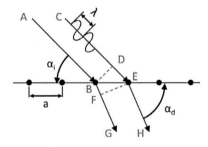

Figure 10.3: Schematic of diffraction of light by a grating with a spacing a.

The difference of the path lengths of the two light beams ABFG and CDEH is $\overline{DE}$ - $\overline{BF}$. If this past length difference is equal to a multiple of the light's wavelength, the light gets constructively diffracted. Depending on the incident angle, α_i, and the grating spacing, a, constructive diffraction takes place at distinct diffracting angles, α_d, as described in the following equation:

$$\left(\overline{DE}-\overline{BF}\right) = a\left(\cos\alpha_i - \cos\alpha_d\right) = n\lambda \tag{10.4}$$

The three Laue equations for x-rays in a 3-dimensional crystal are

$$a\left(\cos\alpha_i - \cos\alpha_d\right) = h\lambda \tag{10.5}$$

$$b\left(\cos\beta_i - \cos\beta_d\right) = k\lambda \tag{10.6}$$

$$c\left(\cos\gamma_i - \cos\gamma_d\right) = l\lambda \tag{10.7}$$

with α_i, β_i, γ_i, and α_d, β_d, γ_d angles between the incident and diffracted x-ray beams, respectively, of the lattice parameters a, b, and c, corresponding to the 1-dimensional grating in Figure 10.3.

10.5.1.3. BRAGG CONDITION

William Lawrence Bragg presented in 1912 the relationship for diffraction between lattice spacing, wavelength and diffraction angle that later became Bragg's law. Constructive interference, e.g. in x-ray, neutron, electron diffraction, follow this well-known Bragg law. It states that for a set of parallel crystal planes, hkl, with a spacing of d_{hkl} and for a wavelength λ (and multiples of it) only certain equal incoming and diffraction angles will lead to constructive interference. The Bragg law is given in the following equation (10.8) and different approach to understand the Bragg law is shown in a schematic in Figure 10.4:

$$n\lambda = 2d_{hkl}\sin\theta \tag{10.8}$$

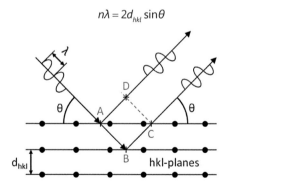

Figure 10.4: Schematic of the relation between wavelength, incident and diffracted beam, and d-spacing described by Bragg. Only if the difference between $\overline{AD}$ and the sum of the distances $\overline{AB}$ and $\overline{BC}$ is equal to the wavelength or a multiple of the wavelength, constructive interference takes place.

10.6. Appendix D: Publications of the Author

Parts of this dissertation have been already published in journals (J) or conference proceedings (C). The number in the square parenthesis is the citation number used in this dissertation.

C5   M.Chmielus, V.A.Chernenko, A.Hilger, G.Kostorz, P.Müllner, R.Schneider, "Magneto-
[38]   mechanical Properties and Fracture of a Mechanically Constrained Ni-Mn-Ga Single
       Crystal After Extended Magnetic Cycling", TMS Special Topic in: G.B. Olson, D.S.
       Lieberman, A.B. Saxena (Eds.), Proceedings of ICOMAT 2008, Santa Fe, NM, June 6–July 5,
       2008, included in parts in chapter 8

J6   M.Chmielus, K.Rolfs, R.Wimpory, W.Reimers, P.Müllner, R.Schneider, "Effects of surface
[111] roughness and training on the twinning stress of Ni-Mn-Ga magnetic shape-memory
       alloys", Acta Mater, 58, 3952-3962 (2010), included in parts in chapter 7

Other publications of the author that are not part of this dissertation.

J5   J.Guldbakke, M.Chmielus, K.Rolfs, A.Raatz, P.Müllner, R.Schneider, J. Hesselbach,
[31]   "Magnetic, Mechanical, and Fatigue Properties of a Ni45.4Mn29.1Ga21.6Fe3.9 single-
       crystalline sample", Scripta Mater, 62, 875-878 (2010)

J4   K.Rolfs, M.Chmielus, R.C.Wimpory, A.Mecklenburg, P.Müllner, R.Schneider, "Double
[30]   twinning in Ni-Mn-Ga-Co", Acta Mater, 58, 7 (2010)

C7   P.Müllner, X.Zhang, Y.Boonyongmaneerat, C.Witherspoon, M.Chmielus, D.C.Dunand,
[25]   "Recent Developments in Ni-Mn-Ga Foam Research", Materials Science Forum, 635, 119-
       124 (2010)

J3   M.Chmielus, X.X.Zhang, C.Witherspoon, D.C.Dunand, P.Müllner, "Giant magnetic-field-
[24]   induced strains in polycrystalline Ni-Mn-Ga foams", Nat Mater, 8, 863-866 (2009)

C6   V.A.Chernenko, K.Oikawa, M.Chmielus, S.Besseghini, E.Villa, F.Albertini, L.Righi, A.Paoluzi,
[117] P.Müllner, R.Kainuma, K.Ishida, "Properties of Co-alloyed Ni-Fe-Ga Ferromagnetic Shape
       Memory Alloys", J Mat Eng Perform, 18, 5-6 (2009)

J2   V.A.Chernenko, M.Chmielus, P.Müllner, "Large magnetic-field-induced strains in Ni-Mn-Ga
[118] nonmodulated martensite", Appl Phys Letters, 95, 104103 (2009)

C4   D.Carpenter, M.Chmielus, A.Rothenbühler, R.Schneider, P.Müllner, "Application of
[119] Ferromagnetic Shape Memory Alloys in Power Generation Devices", TMS Special Topic in:
       G.B. Olson, D.S. Lieberman, A.B. Saxena (Eds.), Proceedings of ICOMAT 2008, Santa Fe, NM,
       June 6–July 5, 2008

C3 [37] M.Chmielus, D.Carpenter, A.Geleynse, M.Hagler, R.Schneider, P.Müllner, "Numerical Simulation of Twin-Twin Interaction in Magnetic Shape-Memory Alloys", Mater Res Soc Symp Proc, 1090, 1090-Z05-26 (2008)

C2 [36] P.Müllner, A.S.Geleynse, D.R.Carpenter, M.S.Hagler, M.Chmielus, "Modeling magnetoelasticity and magnetoplasticity with disconnections and dislocations", Mater Res Soc Symp Proc, 1050, 1050-BB02-01 (2008)

C1 [85] M.Chmielus, V.A.Chernenko, W.B.Knowlton, G.Kostorz, P.Müllner, "Training, constraints, and high-cycle magneto-mechanical properties of Ni-Mn-Ga magnetic shape-memory alloys", Eur Phys J Special Topics, 158, 79-85 (2008)

J1 [23] Y.Boonyongmaneerat, M.Chmielus, D.C.Dunand, P.Müllner, "Increasing Magnetoplasticity in Polycrystalline Ni-Mn-Ga by Reducing Internal Constraints through Porosity", Phys Rev Letters, 99, 247201 (2007)